IS WEAPON SYSTEM COST GROWTH INCREASING?

A Quantitative Assessment of Completed and Ongoing Programs

Obaid Younossi • Mark V. Arena • Robert S. Leonard
Charles Robert Roll, Jr. • Arvind Jain • Jerry M. Sollinger

Prepared for the United States Air Force

RAND PROJECT AIR FORCE

The research described in this report was sponsored by the United States Air Force under Contract FA7014-06-C-0001. Further information may be obtained from the Strategic Planning Division, Directorate of Plans, Hq USAF.

Library of Congress Cataloging-in-Publication Data

Is weapon system cost growth increasing? : a quantitative assessment of completed and ongoing programs / Obaid Younossi ... [et al.].
p. cm.
ISBN 978-0-8330-4135-7 (pbk. : alk. paper)
1. United States—Armed Forces—Weapons systems—Costs. I. Younossi, Obaid.

UF503.I47 2007
355.6'212—dc22

2007022049

The RAND Corporation is a nonprofit research organization providing objective analysis and effective solutions that address the challenges facing the public and private sectors around the world. RAND's publications do not necessarily reflect the opinions of its research clients and sponsors.

Cover design by Stephen Bloodsworth

Published 2007 by the RAND Corporation
1776 Main Street, P.O. Box 2138, Santa Monica, CA 90407-2138
1200 South Hayes Street, Arlington, VA 22202-5050
4570 Fifth Avenue, Suite 600, Pittsburgh, PA 15213-2665
RAND URL: http://www.rand.org/
To order RAND documents or to obtain additional information, contact
Distribution Services: Telephone: (310) 451-7002;
Fax: (310) 451-6915; Email: order@rand.org

Preface

This monograph is part of a RAND Project AIR FORCE project titled "The Cost of Future Military Aircraft: Historical Cost Estimating Relationships and Cost Reduction Initiatives." The purpose of the project is to improve the tools used to estimate the costs of future weapon systems. It focuses on how recent technical, management, and government policy changes affect overall cost. This monograph builds on another document from this study, *Historical Cost Growth of Completed Weapon System Programs* (Arena, Leonard, et al., 2006), and quantifies the magnitude of historical cost growth of weapon systems and determines whether there is a trend. It should interest those involved with the acquisition of systems for the U.S. Department of Defense (DoD) and those involved in the field of cost estimation. The data collection and analysis in this monograph were completed in fall 2005 and thereby include data up to the 2004 Selected Acquisition Reports (SARs). Further, we included only ongoing programs that had matured five years past their respective milestone Bs; this analysis excludes ongoing programs, such as Joint Strike Fighter (JSF), Global Hawk, advanced extremely high-frequency (AEHF) satellite, wideband gapfiller satellite (WGS), National Polar-Orbiting Operational Environmental Satellite System (NPOESS), C-130 avionics modernization program, and C-5 reliability enhancement and reengineering program.

The research reported here was sponsored by Lt Gen Donald J. Hoffman, Principal Deputy, Office of the Assistant Secretary of the Air Force, Acquisition (SAF/AQ), and Blaise J. Durante, Office of the Assistant Secretary of the Air Force, Acquisition Integration (SAF/

AQX), and conducted within the Resource Management Program of RAND Project AIR FORCE. The study's technical monitor was Jay Jordan, technical director of the Air Force Cost Analysis Agency.

Other RAND Project AIR FORCE documents that address weapon system acquisition and cost-estimating issues include the following:

- *An Overview of Acquisition Reform Cost Savings Estimates* (MR-1329-AF), by Mark A. Lorell and John C. Graser.
- *Military Airframe Acquisition Costs: The Effects of Lean Manufacturing* (MR-1325-AF), by Cynthia R. Cook and John C. Graser.
- *Military Airframe Costs: The Effects of Advanced Materials and Manufacturing Processes* (MR-1370-AF), by Obaid Younossi, Michael Kennedy, and John C. Graser.
- *Military Jet Engine Acquisition: Technology Basics and Cost-Estimating Methodology* (MR-1596-AF), by Obaid Younossi, Mark V. Arena, Richard M. Moore, Mark A. Lorell, Joanna Mason, and John C. Graser.
- *Test and Evaluation Trends and Costs in Aircraft and Guided Weapons* (MG-109-AF), by Bernard Fox, Michael Boito, John C. Graser, and Obaid Younossi.
- *Software Cost Estimation and Sizing Methods: Issues and Guidelines* (MG-269-AF), by Shari Lawrence Pfleeger, Felicia Wu, and Rosalind Lewis.
- *Lessons Learned from the F/A-22 and F/A-18E/F Development Programs* (MG-276-AF), by Obaid Younossi, David E. Stem, Mark A. Lorell, and Frances M. Lussier.
- *Price-Based Acquisition: Issues and Challenges for Defense Department Procurement of Weapon Systems* (MG-337-AF), by Mark A. Lorell, John C. Graser, and Cynthia R. Cook.
- *Impossible Certainty: Cost Risk Analysis for Air Force Systems* (MG-415-AF), by Mark V. Arena, Obaid Younossi, Lionel A. Galway, Bernard Fox, John C. Graser, Jerry M. Sollinger, Felicia Wu, and Carolyn Wong.

- *Systems Engineering and Program Management: Trends and Costs for Aircraft and Guided Weapons Programs* (MG-413-AF), by David E. Stem, Michael Boito, and Obaid Younossi.
- *Evolutionary Acquisition: Implementation Challenges for Defense Space Programs* (MG-431-AF), by Mark A. Lorell, Julia F. Lowell, and Obaid Younossi.
- *Historical Cost Growth of Completed Weapon System Programs* (TR-343-AF), by Mark V. Arena, Robert S. Leonard, Sheila E. Murray, and Obaid Younossi.

RAND Project AIR FORCE

RAND Project AIR FORCE (PAF), a division of the RAND Corporation, is the U.S. Air Force's federally funded research and development center for studies and analyses. PAF provides the Air Force with independent analyses of policy alternatives affecting the development, employment, combat readiness, and support of current and future aerospace forces. Research is conducted in four programs: Aerospace Force Development; Manpower, Personnel, and Training; Resource Management; and Strategy and Doctrine.

Additional information about PAF is available on our Web site: http://www.rand.org/paf/

Contents

APPENDIXES

Figures

Tables

Summary

Cost growth in DoD acquisition programs has been a long-standing concern of senior policymakers and members of Congress. In recent decades, there have been numerous attempts to rein in this growth. Some changes involve reforms to the acquisition process, while others entail legislation. The RAND Corporation has a long history of studying cost growth in defense acquisition, with research reaching back to the 1950s.

Research Focus and Approach

The U.S. Air Force asked RAND to examine weapon system cost growth. To do so, we attempted to answer two questions:

- What is the cost growth of DoD weapon systems?
- What has been the trend of cost growth over the past three decades?

To answer the first question, we drew on a recent RAND report on the analysis of cost growth of completed programs (Arena, Leonard, et al., 2006) (see pp. 15–18). To answer the second question, we performed a new analysis of both completed and ongoing programs (see pp. 19–40). The data analyzed came from a SAR database maintained by RAND since the early 1990s (see pp. 9–14).

Cost growth is defined as the ratio between the most recent SAR estimate (or the estimate reported in the program's final SAR) and the cost estimate baseline reported in a prior SAR issued at the time of a given milestone. The values reported in SARs reflect the official position of the management authority of the program—either the Office of the Secretary of Defense (OSD) or one of the military services.

To address the magnitude of cost growth, the first part of this analysis relies on previous RAND work on cost growth analysis of completed programs (specifically, Arena, Leonard, et al., 2006) (see pp. 19–24). But to evaluate the trend over time, we also analyzed some ongoing programs (see pp. 25–29). To measure growth on an equivalent basis with completed programs, we measured cost growth five years after milestone (MS) B for all programs—completed and ongoing (see pp. 31–33).[1] Changes in the mix of system types over time were also considered because earlier studies have suggested that cost growth varies by program type (see pp. 33–35). DoD procures more space and electronics systems and fewer aircraft and helicopters today than it did a decade or two ago.

What Is the Cost Growth of DoD Weapon Systems?

Figure S.1 shows the results of the analysis with respect to the first question. It presents the results of the analysis of 46 completed programs for total development and procurement cost growth, both as a simple and a dollar-weighted average. As the figure shows, the average total cost growth ratio across all programs is 1.46. In other words, on the average, programs cost 46 percent more than estimated at MS B.

---

[1] In this monograph, we use the current acquisition process' milestone designations, i.e., MS A, MS B, and MS C as defined in DoD 5000.1 (U.S. Department of Defense, 2000). (See Chapter One for a full discussion.) However, some of the older programs use the older system of MS I, MS II, and MS III. For simplicity of presentation, we assume that the milestones are directly comparable, e.g., that MS II and MS B are essentially equivalent points in a program.

Figure S.1
Weapon System Cost Growth by Acquisition Phase for Completed Programs

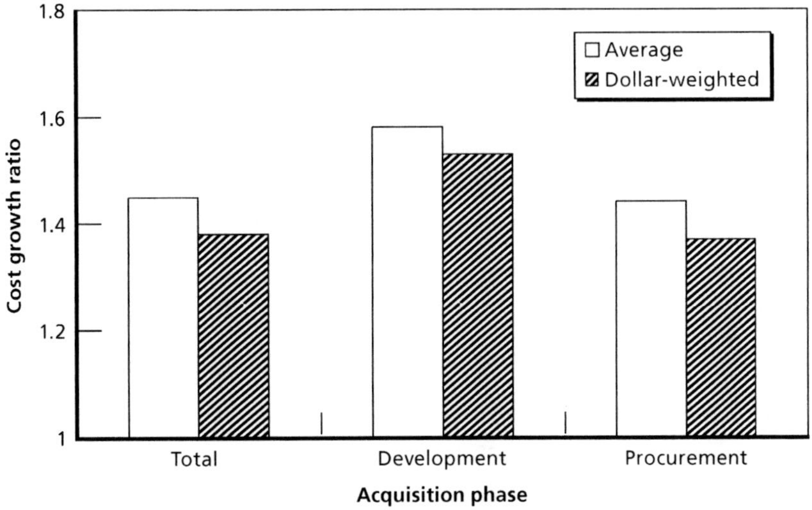

NOTE: All cost growth ratio calculations include adjustments to remove the effects of inflation. Procurement and total cost growth ratio calculations include additional adjustment to remove the effects of changes in the number of systems built in the program's procurement phase.
RAND *MG588-S.1*

The highest cost growth occurs in the development phase, with the ratio reaching almost 1.6. When calculated on a dollar-weighted basis, the growth is slightly less but still substantial.

What Has Been the Trend of Cost Growth?

As Figure S.1 shows, DoD has averaged 46 percent total cost growth in procuring its major weapon systems. It has long been aware of this problem and has initiated a variety of acquisition reforms to reduce cost growth (see Lorell and Graser, 2001, and Hanks et al., 2005). The question is, then, have these reforms reduced cost growth over time? The answer lies in the trend of cost growth over time (see pp. 33–35).

Figure S.2 shows the results of PAF analysis of the time trends in cost growth. It presents five views of mean values of development cost growth in three decades—the 1970s, 1980s, and 1990s. The first bar is development cost growth measured at the program completion point, and the other bars indicate measurements at five years after MS B. We focus on development cost growth because it is a good early indicator of total program cost growth and because its growth trend is similar to procurement and total program cost growth (see pp. 36–38).

The first bar for each decade shows the development cost growth factor (DCGF) of completed programs. Categorization into a particular decade is based on the decade in which each program had its MS B (or equivalent) decision. These first bars indicate declining DCGF, from almost 1.8 in the 1970s to just over 1.2 in the 1990s. However, this trend should be regarded with a measure of skepticism, in part because the averaged data for each decade include programs of different lengths (more recent programs are necessarily shorter) and in part because the mix of types of programs that are completed from each decade is quite different. For example, any project completed in the

Figure S.2
Trend of Weapon System Development Cost Growth

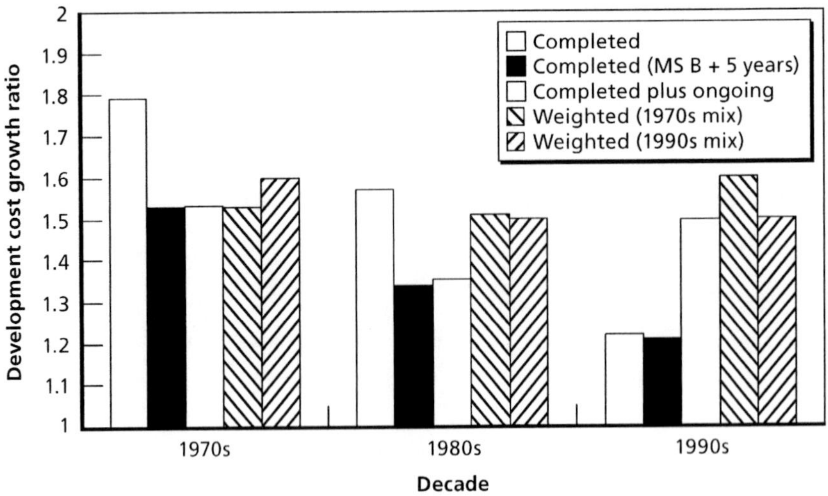

RAND *MG588-S.2*

1990s would have to be of shorter duration and would tend to be of the type that experiences the lowest cost growth (Arena, Leonard, et al., 2006). The previous research shows that electronics programs tend to have less cost growth and that the vast majority of the completed programs in the 1990s fall into that category. The research also shows that shorter projects tend to have less cost growth than do longer ones. Thus, the apparent decline in cost growth may be influenced by the inclusion of shorter projects and commodity classes that typically experience less cost growth (see pp. 38–40).

The second bar in each decade grouping traces the DCGF trend by examining the same set of completed programs, but measured at five years past MS B. As might be expected, measuring the DCGF from the five-year point does result in a lower DCGF. The relative ranking of the decades remains unchanged, but the growth in the 1970s and 1980s is less because all growth after the five-year point is excluded (see pp. 38–40).

The third bar in each decade grouping incorporates 34 ongoing programs with the completed ones, all measured at five years past MS B. The combined sets of programs are compared at this common point. This bar shows that, although DCGF declined between the 1970s and the 1980s, in the 1990s, DCGF rose again to about the same level as in the 1970s. In addition, the large increase in DCGF from the second to third bar in the 1990s grouping indicates that ongoing programs begun in the 1990s have substantially higher DCGFs than do those begun in that decade that are now complete (see pp. 38–40).

Previous research (Arena, Leonard, et al., 2006; McNicol, 2004; Drezner et al., 1993) indicated significant differences among types of weapon systems (e.g., aircraft, missile, electronic). To assess the effect of weapon program type, we controlled for type by normalizing the contribution of each program type to the DCGF based on the proportions of each weapon system type in the 1970s or the 1990s. Chapter Six explains this normalization method. These normalized cost growth figures are shown in the fourth and fifth bars in each decade's grouping in Figure S.2. The fourth bar shows completed and ongoing programs measured at five years past MS B and weighted by the 1970s program mix, and, similarly, the fifth bar shows DCGF of completed and ongoing programs, all measured at five years past MS B and weighted by the 1990s program mix. Controlling for the 1970s mix of programs shows some increase between the 1970s and the 1990s, but not a significant one. Controlling for a 1990s program mix shows a slight decline between the 1970s and the 1990s, but it is not statistically significant. They show that cost growth has not improved over the decades.

We also conducted a statistical analysis to examine whether there were any differences in the trend of development cost growth among various weapon system types and the military services, and we conclude that the three services do not differ significantly in their DCGF levels for their respective programs. We repeated the statistical analysis for development cost growth weighted by development budgets in constant dollars to highlight whether programs with higher development costs have different time trends from those with lower costs. By and large, various weapon system types do not have significantly different trends, with the exception of helicopters (for nonweighted DCGF) and space systems (for weighted DCGF), which did show higher levels of development cost growth than the average of all weapon system types (see pp. 39–40). While the results are statistically significant at the 0.05 level, the small number of observations (eight helicopter programs, three satellite programs, and three launch vehicle programs) included in the analyses means that caution should be used in making any generalizations about these particular results.

Conclusion

Perhaps the most important finding of the analysis is that development cost growth in the past three decades has remained high, with no significant improvement. However, the analysis also suggests that there was greater variability in development cost growth in the 1990s; that is, some observations were substantially higher than the mean. Thus, despite the many acquisition reform and other DoD management initiatives over the years, the development cost growth of military systems has not been reduced. This is not an indictment of the government personnel or contractors involved in the acquisition of systems for the military. There is no doubt that the systems developed in each successive decade are more complex than those of the prior decade. The ever-increasing complexity of technology, software density, system integration complexity, and the like make the estimating a total system's development cost, at the inception of major development activities, an increasingly challenging endeavor (Arena, Blickstein, et al., 2006).

As our very rough comparison with the analysis by Flyvbjerg, Holm, and Buhl (2002) of public works projects shows, weapon system total cost growth is higher than that of rail, fixed-link, and road projects. This difference is understandable given that the technologies involved in those projects are extremely well understood and include the conservative, evolutionary adaptation of new technology over time. Most DoD defense development programs involve much higher levels of new technology adaptation and therefore result in inherently higher levels of cost and schedule uncertainty.

This is also not to say that DoD cannot do better in controlling cost growth. Undoubtedly it can, and it should bend its efforts to doing so. Over the years, several studies, by RAND and others,[2] have attempted to identify the causes of cost growth and what steps can be taken to address them. These causes fall into the following broad areas: overoptimism, estimating errors, unrecognized technical issues, requirements creep, lack of incentives to control cost, and schedule extensions. Therefore, addressing the issue of cost growth requires vigorous involvement of all stakeholders in DoD. However, there may be one aspect of the acquisition process that merits special attention, and that is the cost estimates that form the basis of program budgets. Cost growth may also reflect poor initial budget estimates. Better cost estimates would not necessarily save any money; however, they would provide decisionmakers a better basis for deciding whether to pursue a given program.

---

2 Recent RAND studies that address ways in which DoD can address the growing cost of weapon system development include the following: Arena, Younossi, et al. (2006); Arena, Blickstein, et al. (2006); Younossi, Stem, et al. (2005); Lorell and Graser (2001); and Cook and Graser (2001).

Abbreviations

AABNCP	Advanced Airborne Command Post
AAWS-M	advanced antitank weapon system–medium
ACAT	acquisition category
AEHF	advanced extremely high-frequency [satellite]
AFATDS	Advanced Field Artillery Tactical Data System
AHIP	Army helicopter improvement program
AIM	air interceptor missile
AIWS	advanced interdiction weapon system
ALCM	air-launched cruise missile
AMRAAM	Advanced Medium-Range Air-to-Air Missile
APUC	average procurement unit cost
ATACMS	Army tactical missile system
AWACS	Airborne Warning and Control System
BFVS	Bradley fighting vehicle system
BY	base year
C2	command and control
CAPTOR	encapsulated torpedo

CDR	critical design review
CEC	cooperative engagement capability
CELV	complementary expendable launch vehicle
CGF	cost growth factor
CGS	common ground station
CIC	cost improvement curve
CMUP	Conventional Mission Upgrade Program
CNCE	communication nodal control element
DCGF	development cost growth factor
DMSP	Defense Meteorological Satellite Program
DoD	U.S. Department of Defense
DSCS	Defense Satellite Communication System
DSP	Defense Support Program
DTDMA	distributed time division multiple access
EELV	evolved expendable launch vehicle
FAADS	Forward Area Air Defense System
FCS	Future Combat System
FMTV	family of medium tactical vehicles
FOC	full operating capability
FRP	full-rate production
GAO	U.S. Government Accountability Office
GBS	Global Broadcasting Service
GLCM	ground-launched cruise missile
GMLRS	Guided Multiple Launch Rocket System

GSM	ground station module
HARM	high-speed antiradiation missile
ICBM	intercontinental ballistic missile
ICH	improved cargo helicopter
IOC	initial operating capability
IOT&E	initial operational test and evaluation
IUS	inertial upper stage
JASSM	Joint Air-to-Surface Standoff Missile
JDAM	Joint Direct Attack Munition
JPATS	Joint Primary Air Training System
JSF	Joint Strike Fighter
JSIPS	Joint Service Imagery Processing System
JSOW	Joint Standoff Weapon
JSTARS	Joint Surveillance Target Attack Radar System
JTACMS	Joint Tactical Missile System
JTIDS	Joint Tactical Information Distribution System
LANTIRN	Low-Altitude Navigation and Targeting Infrared System for Night
LOS-R	line of sight–rear
LRIP	low-rate initial production
MCS	Maneuver Control System
MDAP	major defense acquisition program
MIDS-LVT	Multifunctional Information Distribution System–Low-Volume Terminal

MILCON	military construction
MS	milestone
NAS	National Airspace System
NASA	National Aeronautics and Space Administration
NAVSTAR	Navigational Satellite Timing and Ranging
NEACP	National Emergency Airborne Command Post
NPOESS	National Polar-Orbiting Operational Environmental Satellite System
O&M	operation and maintenance
O&S	operations and support
OLS	ordinary least squares
OSD	Office of the Secretary of Defense
OTH-B	over-the-horizon backscatter radar
PAF	Project AIR FORCE
PAUC	program acquisition unit cost
PDR	preliminary design review
PRP	Propulsion Replacement Program
R&D	research and development
RDT&E	research, development, test, and evaluation
RERP	reliability enhancement and reengineering program
RSIP	Radar System Improvement Program
S&T	science and technology
SAF/AQ	Office of the Assistant Secretary of the Air Force, Acquisition

SAF/AQX	Office of the Assistant Secretary of the Air Force, Acquisition Integration
SAR	Selected Acquisition Report
SBIRS-Hi	Space-Based Infrared System–High
SHORAD	short-range air defense
SINCGARS-V	Single-Channel Ground and Airborne Radio System–VHF
SMART-T	Secure Mobile Antijam Reliable Tactical Terminal
T&E	test and evaluation
TDMA	time division multiple access
TJS	Tactical Jamming System
TRI-TAC	joint tactical communications
TY	then-year
UHF	ultrahigh frequency
USMC	U.S. Marine Corps
UTTAS	Utility Tactical Transport Aircraft System
VHF	very high frequency
WGS	wideband gapfiller satellite

CHAPTER ONE

Introduction

Cost growth has been a perennial problem that has affected most major defense weapon systems. RAND Corporation studies going back to the 1950s have researched weapon system cost growth. As early as 1959, Marshall and Meckling noted how inaccurate and biased input parameters influence the program cost estimates:

> "Early" estimates of important parameters are usually quite inaccurate. They are inaccurate in two respects. First such estimates are "biased" toward overoptimism. Second, aside from the bias, the errors in the estimate evidence a substantial variation. (Marshall and Meckling, 1959)

This statement rings true in the context of current U.S. Department of Defense (DoD) discussion about the cost of weapon systems. The issue of inaccurate cost estimates clearly concerns Congress, which has enacted legislation in an attempt to limit the spiraling cost of modern weapon systems. Its most recent effort requires Pentagon officials to compare current unit prices with original cost estimates as well as with more recent, revised projections (see Karp, 2006b). The topic has been studied intensively, and a myriad of RAND reports as well as research by other institutions have developed recommendations on how to control or at least mitigate this cost growth.[1]

---

[1] For more recent analysis, see Arena, Leonard, et al. (2006); McNicol (2004); Tyson, Nelson, et al. (1998); Tyson, Harmon, and Utech (1994); and Drezner et al. (1993).

Background

In an era of limited resources and ever-increasing cost to develop new military systems, the cost growth of weapon systems is, once again, a topic in the press (see Karp, 2006a). Recent reports discuss the coming "bow wave" of weapon costs yet to be fully understood. Congressional testimony by DoD acquisition leaders has focused on plans to curb this trend (Bruno, 2005; Karp and Pasztor, 2005; Sevastopulo, 2005).

A recent U.S. Government Accountability Office (GAO) study examined 54 ongoing DoD acquisition programs in various stages of development (U.S. Government Accountability Office, 2005). The programs examined in that study represented around $800 billion in DoD investment; came from all the defense agencies and military services; and included aircraft, missiles, space systems, ground vehicles, unmanned aerial vehicles, and information technology. Table 1.1 shows four major programs that have experienced either substantial cost growth or quantity reductions.

Since weapon system progress is always described in terms of DoD acquisition milestones, we briefly review them in the next section.

Table 1.1
Comparisons of Four Weapon Systems' Initial Cost Estimates and Quantities to the Latest Cost Estimates and Quantities

System	Initial Cost Estimate ($ billions)	Most Recent Cost Estimate ($ billions)	Initial Quantity	Latest Quantity
Joint Strike Fighter (JSF)	186.6	198.6	2,866	2,457
Space-Based Infrared System–High (SBIRS-Hi)	3.9	9.9	5	5
F/A-22	78.9	73.1	648	279
Future Combat System (FCS)	79.8	108.0	15	15

SOURCE: U.S. Government Accountability Office (2005).

DoD Weapon System Development Major Milestones

Most DoD weapon systems follow guidelines set forth Department of Defense Instruction 5000.2, signed on May 12, 2003 (U.S. Department of Defense, 2003). This instruction divides the defense acquisition management process into five separate phases, as illustrated in Figure 1.1. The three lettered triangles along the top of the boxes represent key decision points, or milestones, and are commonly referred to by their letter designations.

Prior to engaging in formal system development, as shown in Figure 1.1, DoD funds various types of science and technology (S&T) activities aimed at basic and applied research. These projects are not weapon system–specific and none of their costs is directly associated with any weapon system. The first phase of this formal DoD system is called *concept refinement*, and its purpose is to refine the initial concepts and lay out a system development strategy. During this phase, alternative system concepts are evaluated. These concepts will consider functionalities ranging from commercial off-the-shelf through

Figure 1.1
Major Weapon System Phases and Milestones

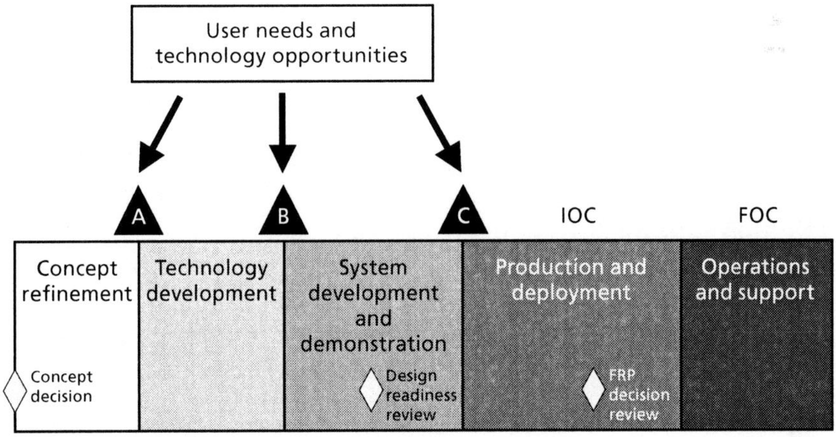

SOURCE: U.S. Department of Defense (2003).
NOTE: IOC = initial operating capacity. FOC = full operating capacity. FRP = full-rate production.
RAND *MG588-1.1*

promising S&T solutions to achieve specific military needs. It is followed by the *technology development* phase, in which the focus is on reducing technical risk and determining the appropriate set of technologies to be integrated into a full system. The decision point at which to enter the technology development phase is milestone A (represented by the first triangle in Figure 1.1). At this and subsequent milestones, decisionmakers assess user needs and technology opportunities and approve the transition to the next phase of acquisition.

The next phase is *system development and demonstration*. The decision point is milestone (MS) B (represented by the second triangle in Figure 1.1). The purposes of this phase are to develop a new capability, mitigate integration and manufacturing risk, ensure that the system can be supported, consider human interface issues, ensure that the system is cost-effective, and ensure that all critical technologies are adequately protected. Furthermore, this phase demonstrates system integration, interoperability of the system among services and allies, and safety mechanisms.

At MS C (represented by the third triangle in Figure 1.1), the decisionmaking authority approves the program for production and deployment. The production phase includes low-rate initial production (LRIP) and FRP. During the LRIP phase, small quantities are approved for production mainly for the purpose of completing the operational testing requirements and demonstrating the integration of various onboard weapon systems, and an IOC is established. FRP results in higher quantities to meet the military's operational needs.

The final phase of a weapon system's life cycle is the *operations and support* (O&S) phase, which occurs after FOC is established. The O&S phase is divided into two subphases: sustainment and disposal. The sustainment phase includes all functions related to maintenance, transportation, staffing, configuration control, data management, and so on. At the end of the weapon system's useful life, it is demilitarized and disposed of in accordance with safety, security, and environmental regulations.

Research Questions

To examine weapon system cost growth, we explore two main questions:

- What is the cost growth of DoD weapon systems?
- What has been the trend of cost growth over the past three decades?

We define *cost growth* as the ratio between the most recent or final value given in the Selected Acquisition Report (SAR) and the cost provided in the SAR at some earlier milestone. These values do not reflect any individual organization's cost estimate but, rather, what the military services asked for in their budget requests and what Congress subsequently authorized and appropriated.

To answer the first question, we draw on a recent RAND report (Arena, Leonard, et al., 2006) that presents an analysis of cost growth of completed programs. To answer the second question, we perform a new analysis of both completed and ongoing programs. The data we used came from an SAR database that RAND has maintained since the early 1990s. (See Chapter Two for further discussion of SARs.)

Research Approach

Typical cost growth analyses include programs at various stages of the life cycle (McNicol, 2004; Drezner et al., 1993). For example, some included programs may be in the very early stage of development, whereas some other programs might be fairly mature or perhaps even finished. Programs early in development may or may not experience much cost growth because technical problems or issues have yet to surface. However, mature programs are likely to have experienced most of the growth that will occur by the time of their completion and thus provide a fairly complete picture of the magnitude of the cost growth.

To address the magnitude of cost growth, the first part of this analysis relies on previous RAND work on cost growth analysis of completed programs (Arena, Leonard, et al., 2006). We summarize the findings of that research in Chapter Two. To evaluate cost growth of programs begun at different points in the past three decades and to include more recently begun programs that are of primary importance to DoD and Congress, we also analyze some ongoing programs. We added a sample of ongoing programs so that we could examine growth trends over the past few years. As mentioned earlier, ongoing programs tend to exhibit lower cost growth than do completed ones. To measure growth on an equivalent basis, we measure development cost growth at five years after MS B.[2] We also adjust cost growth factors (CGFs) to account for changes in the proportion of each program type (e.g., aircraft, electronics, missile) begun in the past three decades. For example, we adjust for the fact that DoD is in the process of acquiring more space systems and electronics programs today than it was a decade or two ago. As was shown in Arena, Leonard, et al. (2006), cost growth varies by program type. Moreover, we perform statistical analyses to examine the development cost growth differences by various weapon system type and by military service.

Organization of This Monograph

This monograph is organized in the following way. Chapter Two describes the source of our data and methodology for this analysis. Chapter Three summarizes historical cost growth of completed weapon system programs. Chapter Four examines the completed programs' development cost growth trend over time. Chapter Five compares this trend of development cost growth in completed programs with trends in ongoing aircraft, space, and missile programs. Chapter Six incorporates all ongoing programs five years beyond MS B in a development cost growth trend analysis. Chapter Seven summarizes other megaprojects and their cost growth compared to the weapon system cost growth. Finally, Chapter Eight presents some concluding remarks.

---

[2] In this monograph, we use the current acquisition process' milestone designations, i.e., MS A, MS B, and MS C. However, some of the older programs use the older system of MS I, MS II, and MS III. For simplicity of presentation, we assume that the milestones are directly comparable, e.g., that MS II and MS B are essentially equivalent points in a program.

This monograph includes four appendixes. Appendix A presents a short description of all weapon system programs included in the analysis of completed program cost growth. Appendix B includes all programs that are still ongoing but that are at least five years past their MS B decision points. Appendix C includes statistical and sensitivity analyses of development cost growth trends, and Appendix D presents a statistical comparison of development cost growth among the military services.

CHAPTER TWO

Data and Methodology

This chapter describes the source of our data and our methodology for this analysis.

Selected Acquisition Reports

The data used in this analysis are derived from selected acquisition reports, or SARs. These reports are required by a federal law (see 10 USC 2432) and are prepared at least annually for all major defense acquisition programs (MDAPs) by the program office responsible for each program's execution.[1] Only the largest (in dollar terms) of programs are required to submit a SAR: those that exceed a $365 million development cost threshold or $2,190 million procurement threshold.[2] Programs that are not this large but are designated as "special interest" by Congress must also submit SARs. In a typical year, the value of the SAR programs in aggregate represent 80 percent of the value of DoD's entire acquisition portfolio; thus, these reports provide Congress with the status of much of ongoing military acquisition. The SAR provides (to the maximum extent possible) standardized information to Congress annually.

---

1 MDAPs are those designated as such by either acquisition category (ACAT) 1C or ACAT 1D.

2 These threshold values are expressed in FY 2000 dollars.

SARs provide top-level program descriptions and status information, as well as contact data and source-of-funding information.[3] The program executive summary section includes a narrative of program history and details of recent program events. SARs include cost, schedule, and performance baseline estimates and current estimates for the program. They identify whether a program has "threshold breaches" in any of these key metrics.[4] SARs also summarize current estimates for selected historical and all future scheduled milestones. Cost estimates are typically broken out among research and development (R&D), procurement, military construction, acquisition-related operation and maintenance (O&M),[5] and support. The SAR also explains any changes that occurred since the last report, such as changes to the major milestone dates or key performance metrics. Other data provided include annual funding and procurement quantity, explanations of cost changes from the program's prior SAR, status of the program's major contracts, planned and actual quantity delivered, and expenditures to date.

The most detailed data in the SAR are the cost variance analyses. Each program is required to report the reasons for any cost growth. These are assigned to seven variance categories: economic, quantity, schedule, engineering, estimating, support, and other. The definitions of these categories are sometimes inconsistent; for instance, a reason for cost growth in one program might be categorized under *engineering*, whereas another program might list the same reason under *schedule*. Thus, the cost variance section is not particularly useful when comparing sources of cost growth among programs.

---

3 The program elements and the service-specific appropriation categories that fund the program are identified.

4 Threshold breaches entail the failure to meet plans by specified margins. As an example, a six-month slip precipitates a breach in a scheduled milestone event. A cost increase in excess of 15 percent is considered a breach, with more involved explanations and remediation required when higher percentage increases occur.

5 Some programs use O&M funds to accomplish acquisition-related tasks, such as installation of equipment items.

RAND SAR Database

RAND Project AIR FORCE maintains a database of SAR data going back to the very first SARs submitted to Congress in 1968. The following data are collected and are updated as needed:

- program name (past names are noted)
- service (joint participants are noted)
- prime contractor (not updated after down-select, which typically occurs at MS B or its equivalent)
- prototype (built as part of the acquisition strategy)
- prototyping phase (if applicable)
- modification (if the program modifies an existing system or builds a modified version of an existing design)
- MS 2 or MS B currently planned or actual date
- MS 3a, MS 3, or MS C currently planned or actual date
- initial operational delivery currently planned or actual date
- initial operational test and evaluation (IOT&E) start, currently planned, or actual date
- IOT&E complete currently planned or actual date
- development costs in millions of base-year (BY) dollars and then-year (TY) dollars for the entire program
- procurement costs in millions of BY and TY dollars for the entire program
- military construction (MILCON) costs in millions of BY and TY dollars for the entire program
- O&M costs in millions of BY and TY dollars for the entire program
- R&D quantity
- procurement quantity
- annual funding by acquisition appropriation category for the program's entire duration.

The database includes around 220 programs from all military services. SAR data for many of the programs were not appropriate for use in this analysis. In quantifying the magnitude of cost growth, we included only programs with a "solid" MS B or MS C baseline estimate and that were not subsequently terminated. The criterion for a solid baseline estimate was that it be made within about one year of the date of commitment to that program phase.[6] For the completed sample, we considered programs that were 90-percent complete, measured in either planned program expenditures or total quantity delivered. Congress drops the SAR requirement when programs have met either of the 90-percent completion criteria.

The other inclusion criterion was that only programs of similar type to those systems acquired by the U.S. Air Force would be considered. This requirement resulted in the exclusion, for example, of ships and submarines. Of all the programs in the database, 68 completed programs met all criteria. The list of 68 programs and a narrative describing the source of information for each program is included in Appendix A. Only 46 of the 68 completed programs had an MS B estimate.

In our analysis incorporating more recent system acquisitions, we add ongoing programs that were at least five years beyond MS B. This added 33 programs to the analysis. The list and narrative description of these programs is in Appendix B.

The data in all programs, completed and ongoing, were normalized for the effect of inflation and changes in production quantity from the reference milestone.

Limitations of SAR Data

Using SAR data to study cost growth has several limitations (Hough, 1992). First, the most detailed cost data contained in the SARs are at the levels of annual development funding, procurement funding, MILCON funding, and O&M funding; thus, analysis of cost growth by major subsystems is not feasible. Second, program baselines and

[6] Many programs do not have baseline estimates coincident with MS B or MS C because, prior to the mid-1980s, the criteria for requiring SARs from MDAP-sized programs were less stringent. As a result, the initial SARs for many programs were prepared and submitted to Congress two or more years after the corresponding milestone date.

system configurations evolve over time, complicating the analysis of cost growth. Third, reporting requirements and report preparation guidelines, as well as dollar reporting thresholds, have changed; therefore, comparison across time periods is challenging. Fourth, allocation of cost growth in each of the SAR cost variance categories is inconsistent among programs. Fifth, not all programs are required to provide SARs; classified and special-access programs most likely do not generate SARs. Sixth, and potentially most important, the initial estimates in the SAR are consistent with the program's initial budget and may not reflect the most realistic cost estimate. Whether the baseline values represent the cost estimate developed by the program manager, contractor proposal, independent cost-estimating agency, service, or OSD budget position is an open question. Therefore, anyone using the SAR data to analyze weapon system cost growth should keep these limitations in mind.

Quantity Adjustments

One issue in evaluating cost growth is adjusting for procurement quantities that change over the acquisition life cycle of programs. Sometimes production quantity declines, and other times it increases. The F-22 program at MS B had an estimated production quantity of 648 aircraft but planned only 172 as of the December 2004 SAR. The Joint Direct Attack Munition (JDAM) program at MS B had an estimated production quantity of 87,946 guidance kits, but planned some 211,515 in its December 2004 SAR. Regardless of the direction of a program's quantity change, it is necessary to adjust for this change when analyzing growth from baseline cost estimates. We do so by using the annual funding and quantity data provided in the SARs. From these data, we develop a cost improvement curve (CIC) to rationalize the quantity actually procured with that of the baseline estimate. On rare occasions, we cannot determine a CIC because there are too few units in the program. This often occurs in satellite programs. In these cases, we infer a CIC from the experience of similar programs.

There are two common methods of adjusting cost growth for quantity changes. The first uses the CIC data defined by the annual data for the baseline estimate associated with the reference milestone. This is used to adjust the baseline estimate procurement costs from the baseline's estimated quantity to the program's final quantity. If the actual procurement quantity increases from that anticipated at the reference milestone, the procurement cost estimate at the reference milestone is increased. If the actual procurement quantity decreases from that anticipated at the reference milestone, the procurement cost estimate at the reference milestone is decreased.

The second method uses the CIC data defined by the current, most recent, or final procurement cost data. This CIC is used to adjust the final procurement estimate's quantity to that in the baseline estimate. If the actual procurement quantity increases from that anticipated at the reference milestone, the final procurement cost estimate is decreased. If the actual procurement quantity decreases from that anticipated at the reference milestone, the final procurement cost estimate is increased.

In the analysis reported in Chapters Three and Four, we adjust the baseline estimate to the final program quantity. In practice, either method yields similar results when developing aggregate statistics from a broad sample of data (see Arena, Leonard, et al., 2006).

CHAPTER THREE

Cost Growth of Completed Programs

This chapter summarizes findings originally presented in a prior RAND report, *Historical Cost Growth of Completed Weapon System Programs*, by Mark V. Arena, Robert S. Leonard, Sheila E. Murray, and Obaid Younossi.

The primary growth metric is the CGF, which is the ratio of a program's final cost to its estimated costs (using either MS B or MS C estimates).[1] The results of an analysis of completed program cost growth showed that, by and large, DoD and the military departments have underestimated the budget required for developing and procuring new weapon systems. This analysis included data from 46 weapon system programs that met the criteria of having delivered 90 percent of their final quantities and had provided a cost estimate at the MS B decision. Figure 3.1 is a frequency plot of the total CGF for quantity-adjusted growth reflecting the estimate at MS B. The bars in the figure show the frequency distribution and the line represents the lognormal fit of the data. As the figure shows, the vast majority of programs experience cost growth.

Table 3.1 presents a compilation of simple statistics of the total CGF, development CGF (DCGF), and procurement CGF for completed weapon system programs at MS B. Quantity-adjusted and unadjusted values are shown for both total and procurement growth.

---

[1] The previous study (Arena, Leonard, et al., 2006) used the older milestone designations of MS I, MS II, and MS III. For this summary, we have redesignated the relevant milestone points (i.e., MS II becomes MS B and MS III becomes MS C) to be consistent with the current milestone designations.

Figure 3.1
Distribution of Total Cost Growth from MS B Adjusted for Production Quantity Changes

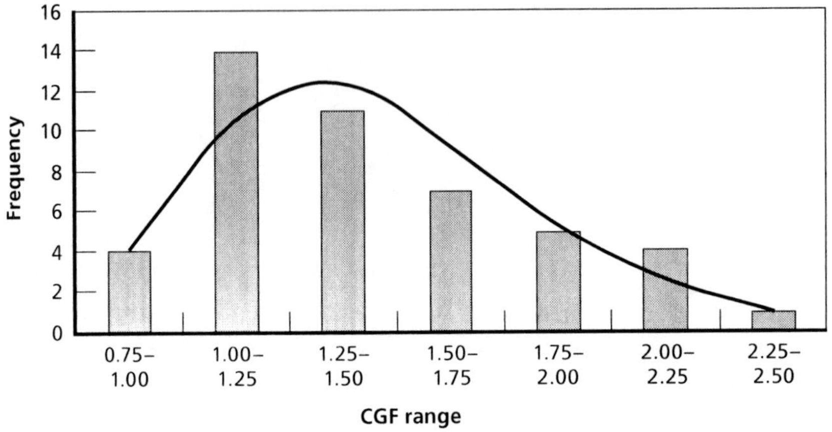

SOURCE: Arena, Leonard, et al. (2006).
RAND *MG588-3.1*

Table 3.1
CGF Summary Statistics by Funding Categories at MS B

Category	Observations	Mean	Median	Standard Deviation	Min.	Max.
Total (adjusted)	46	1.46	1.44	0.38	0.77	2.30
Total (unadjusted)	46	1.65	1.25	1.08	0.37	5.56
Development	46	1.58	1.34	0.79	0.77	5.47
Procurement (adjusted)	44	1.44	1.40	0.42	0.51	2.29
Procurement (unadjusted)	44	1.73	1.30	1.37	0.28	7.28
MILCON	10	1.33	1.11	0.82	0.51	2.87

SOURCE: Arena, Leonard, et al. (2006).

The analysis indicates systematic underfunding of weapon system program costs. Our analysis of the data indicates that the average adjusted total cost growth for a completed program was 46 percent from MS B and 16 percent from MS C. In addition, this analysis showed that programs with longer duration had greater cost growth. Further, as shown in Table 3.2, electronics programs exhibited the lowest cost growth among various weapon systems, and that difference is statistically significant. Moreover, the space programs (satellites and launch vehicles) also had significantly higher development costs than did other weapon systems, but that difference is driven by Titan IV. In addition, we found no statistically significant differences among the military services' average CGFs.

Table 3.2
CGF for Adjusted Total Cost by Commodity Class at MS B

Commodity	Mean	Standard Deviation	Observations
Aircraft	1.35	0.24	9
Cruise missiles	1.64	0.40	4
Electronic aircraft	1.52	0.47	5
Electronics	1.23	0.33	12
Helicopters	1.76	0.21	3
Launch vehicles	2.30	N/A	1
Missiles	1.52	0.38	8
Other	1.40	N/A	1
Satellites	1.55	0.57	2
Vehicles	1.67	N/A	1
Weighted average	1.46	0.38	—
Total observations			46

SOURCE: Arena, Leonard, et al. (2006).

Other Findings of the Study

This analysis shows about 0.20 points greater growth than the previous RAND SAR study did (see Drezner et al., 1993). We attribute this increase to the use of only completed programs in the current analysis:

- The distribution of cost growth is approximately lognormal.
- There is no correlation between development and procurement cost growth.
- There is a recent trend to reduce quantities as a means of controlling cost growth.
- We found no statistically significant differences among the three services.

CHAPTER FOUR

Trends in Cost Growth for Completed Programs

Chapter Three showed cost growth in absolute terms. This chapter looks at the cost growth trend over time. First, we examine the trends in cost growth within a program. Next, we examine how the magnitude of cost growth at a specific point in the acquisition life cycle has changed over the past few decades.

Cost Growth Trends Within a Program

To examine the cost growth trend over time using our completed program set, we plot the ratio of the cost for a particular SAR relative to that at MS B as a fraction of the time between the MS B date and the last SAR. In other words, we are showing the "realized" cost growth at a given point in a program's maturity. Figure 4.1 displays development cost growth plotted against the fraction of time between the actual MS B date and the actual date of the last SAR.[1] The solid line is the mean cost growth, and the dashed lines are the 25-percent (bottom) and 75-percent (top) sample intervals.

Figure 4.1 shows that development cost growth increases fairly steeply up to about the 40-percent point between MS B and the last SAR, and then grows more modestly afterwards. However, development cost does continue to grow until the end of procurement. We

[1] We used the actual time required for the program to progress from MS B to its completion (90-percent) point. Most MDAPs experience schedule slips, making any point in this ratio smaller than if the planned schedule were used.

also examined quantity-adjusted procurement costs. These results are shown in Figure 4.2.

Figure 4.1
Development Cost Growth Versus the Fraction of Time Between MS B and the Final SAR

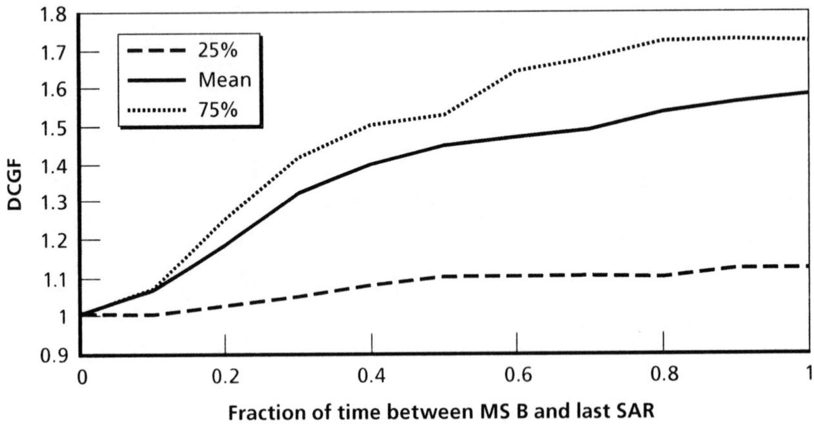

RAND *MG588-4.1*

Figure 4.2
Adjusted Procurement Cost Growth Versus the Fraction of Time Between MS B and the Final SAR

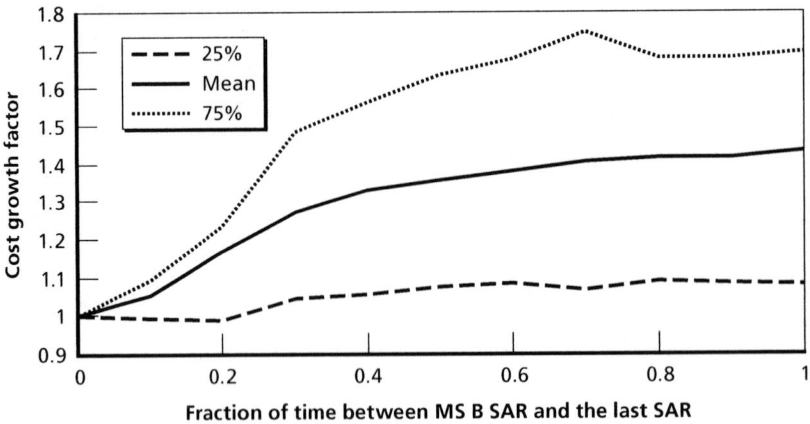

RAND *MG588-4.2*

Similar to development cost, procurement cost continues to grow throughout a program's acquisition phase. As was the case with development cost, procurement cost experiences the most by the 40-percent point in time between its MS B date and the final SAR.

Evolution of Development Cost Growth Over the Past Few Decades

The main question addressed in this monograph is whether cost growth is improving—decreasing—as DoD acquisition processes are evolving. If so, one would see a decreasing trend in average cost growth over time. To examine whether such a trend exists, we calculate the average DCGF for each decade. We categorize each program by the decade in which its MS B occurred. Figure 4.3 shows the results of that analysis.[2]

Figure 4.3
Average DCGF for Each Decade

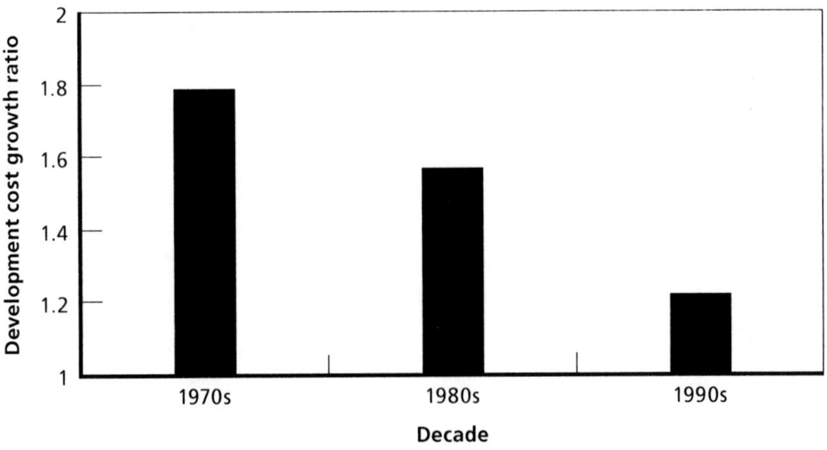

RAND *MG588-4.3*

---

[2] The data include three programs from the 1960s that are not included in Figures 4.3 and 4.4.

The data do indeed show a decline in the cost-growth ratio over time. But any conclusion that a true trend of declining cost growth exists must be regarded with a degree of skepticism. Recall that our sample includes only completed programs. Therefore, any recent programs, i.e., from the 1990s, in our sample would necessarily have short program durations in order to have been completed. One of the important observations we note for completed programs is that programs of shorter duration have lower cost growth. So, the question arises of whether the improvement shown in Figure 4.3 results because completed programs that were initiated in the 1970s may have longer durations than those initiated in the 1990s, and so on.

Figure 4.4 shows the average program duration for the same sample of programs. The duration data also show a declining trend, with a significant decrease in duration for the programs from the 1990s.

Another possible explanation for the decline in cost growth over time is that the mix of program types changed with time. Chapter Three indicated that certain program types, e.g., electronics, had lower-than-average cost growth. Figure 4.5 displays the mix of completed weapon system programs used in the analysis.

Figure 4.4
Average Program Duration Between MS B and the Final SAR

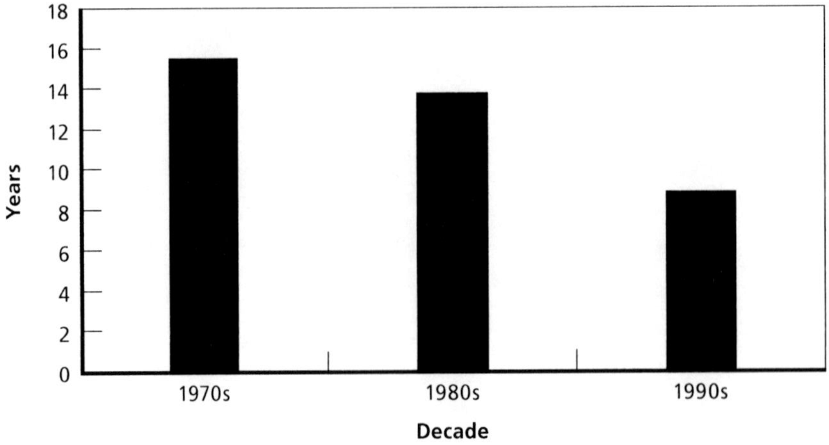

RAND *MG588-4.4*

Figure 4.5
Mix of Completed Programs Categorized by Decade

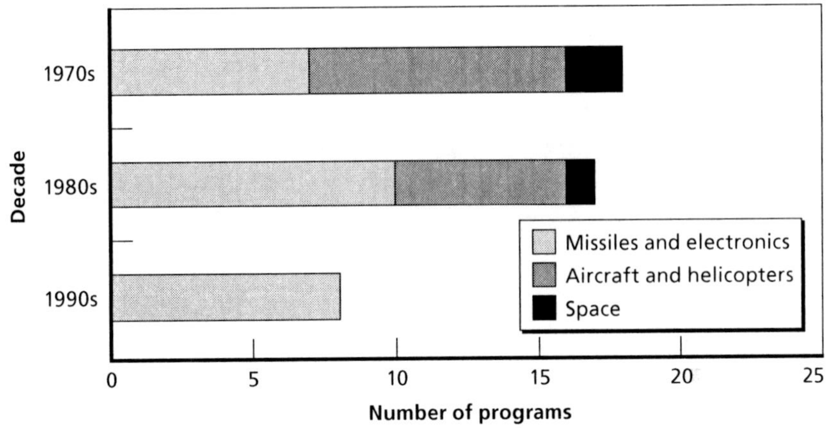

RAND *MG588-4.5*

For the purpose of this analysis, we grouped all types of aircraft and helicopters in a category called *aircraft and helicopters*. Satellites and launch vehicles are grouped in a category called *space*. All the missiles and electronics are grouped in a category called *missiles and electronics*.[3] The mixes of 1970s programs and 1980s programs are relatively comparable. The mix of the 1990s programs differs significantly, since it includes only missiles and electronics. As mentioned earlier in the analysis of completed programs, electronics programs exhibit less cost growth than the average of all programs. Further, many of the aircraft and space programs initiated in the 1990s are still ongoing.

Summary

We have examined the trends in cost growth both over the acquisition life cycle of a program and at program completion using MS B events

3 This grouping approach was motivated by the technical similarities of the weapon system classes. However, to further examine the effect of this grouping, Appendix C provides statistical analyses that determine that, even when we tried different groupings or to assess each individual weapon system type, as a group, the bottom-line conclusion did not change.

that occurred over the past several decades. The trend in development and procurement cost growth over the acquisition life cycle shows continuous growth throughout program execution, though most of the growth occurs in the first half of the program. We observed a decrease in the average cost growth for completed programs from the 1970s through the 1990s. However, this observation appears to be biased by the fact that the more recent programs are of shorter duration and are mostly of the electronics type, both traits that lead to lower cost growth.

CHAPTER FIVE

Cost Growth of Ongoing Programs

This chapter compares the trend of development cost growth of all completed programs with that of ongoing aircraft, space, and missile programs.

Ongoing Aircraft Program Cost Growth

Figure 5.1 compares the trend of development cost growth of completed programs to some major ongoing aircraft development programs. The solid line in the figure is the same as was presented in the Figure 4.1 in Chapter Four and indicates average development cost growth increases over the acquisition life cycle for systems of the types that the Air Force acquires. In the same figure, we plotted development cost growth for selected ongoing programs based on 2004 SAR data (the most recent available at the time of this analysis). As shown in Figure 5.1, the majority of ongoing aircraft development programs are experiencing higher-than-average or average cost growth.[1] Programs in Figure 5.1 that are currently in development are identified by diamonds; those in production are identified by squares.

F-35 or JSF, which is still very early in development, is experiencing significant growth in development costs. Airframe weight increases have led to a redesign of airframe structure (U.S. Government Accountability

[1] Here, we use data from the 2004 SAR to provide a snapshot in time. Historically, most MDAPs experience a schedule slip; therefore, these points may be higher (i.e., cost growth versus time) than where they ultimately may end up.

Figure 5.1
Major Ongoing Aircraft Programs Compared to Trend of Completed Programs

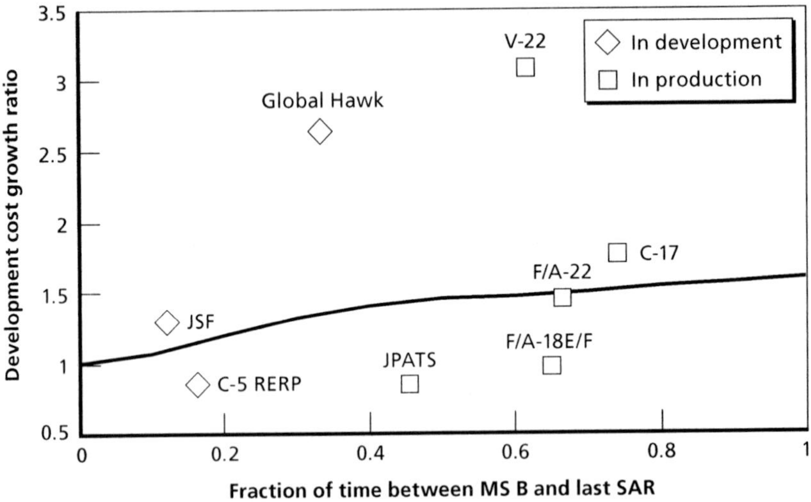

NOTE: JPATS = Joint Primary Air Training System.
RAND *MG588-5.1*

Office, 2005). In 2004, the program reported a Nunn-McCurdy unit cost breach.[2] The V-22 tilt rotor aircraft has been in development since the early 1980s and has yet to achieve IOC. The testing program revealed interoperability and human factor issues that will need to be resolved before FRP (U.S. Government Accountability Office, 2005). The Global Hawk unmanned aerial vehicle program was restructured, which involved an airframe redesign and new sensors. In the F/A-18E/F program, most of the follow-on development effort (after production began) was funded in separate programs; thus, its development cost growth at halfway through production is artificially low.[3]

---

2 "A 'Nunn-McCurdy' unit cost breach occurs when [an MDAP] experiences an increase of at least 15% in Program Acquisition Unit Cost (PAUC) or Average Procurement Unit Cost (APUC) above the unit costs in the Acquisition Program Baseline" (U.S. Department of Defense, 2002, p. 1).

3 F/A-18EF research, development, test, and evaluation (RDT&E) funding reported in the SAR ends in 2002. Most programs have RDT&E funding extending to the end of the production or beyond.

The program also benefited from a low-risk technology development, and its budget was based on realistic independent cost estimates (Younossi, Stem, et al., 2005). JPATS is a commercially derived training aircraft with development and production contracts awarded just days apart. Its development program represents just 6 to 7 percent of its total acquisition value, and its production CGF is greater than 1.5. The C-5 RERP is also plotted in the figure. (Programs in Figure 5.1 that are currently in development are identified by diamonds; those in production are identified by squares.)

Ongoing Space Program Cost Growth

Figure 5.2 displays the same results as Figure 5.1, but here we plot five ongoing space programs for comparison with the historical record of completed programs. The four satellite programs are above the line

Figure 5.2
Major Ongoing Space Programs Compared to Trend of Completed Programs

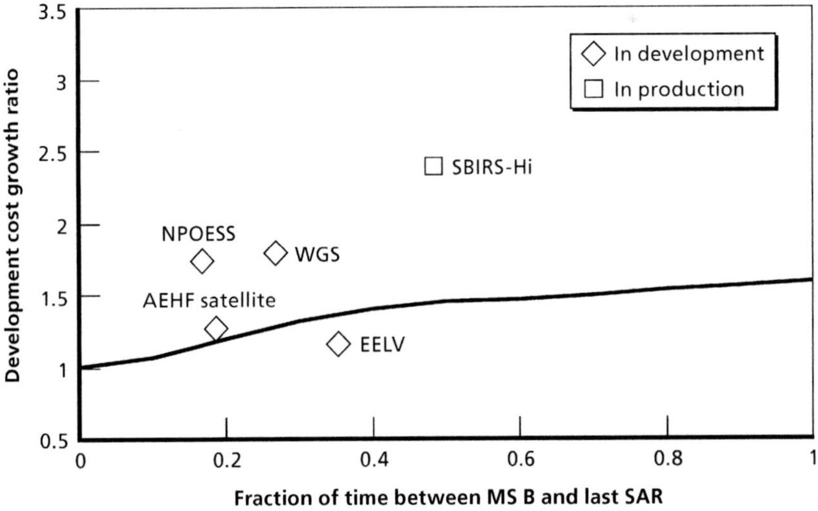

NOTE: NPOESS = National Polar-Orbiting Operational Environmental Satellite System. WGS = wideband gapfiller satellite.

RAND *MG588-5.2*

representing the trend of completed programs cost growth. The evolved expendable launch vehicle (EELV) falls below the line.

The advanced extremely high-frequency (AEHF) satellite is an international joint-venture communication satellite program. The program has experienced schedule delay related to its cryptological equipment and command-post terminal development (U.S. Government Accountability Office, 2005). The AEHF satellite program recently reported a Nunn-McCurdy unit cost breach. The WGS is a joint Air Force–Army program that provides the military with communication services. Manufacturing issues have contributed, in large part, to schedule delays (U.S. Government Accountability Office, 2005). SBIRS-Hi is a space-based missile warning system. In May 2004, the program filed a second Nunn-McCurdy breach. Lately, the SBIRS-Hi program again has captured headlines as a result of its substantial cost growth and schedule issues.

Lastly, the EELV is an industry partnership between the Air Force and commercial launch services. Although the development cost growth for this program falls below the average, no one really knows how much the contractor has spent. The information reported in the program SAR accounts only for the government's investment. Development was treated as if the system were commercially derived; thus, its development and production contracts were signed on the same day. Its development program represents just 6 to 7 percent of the total acquisition value, and its production CGF is greater than 2.1. NPOESS is also shown for comparative purposes.

Ongoing Missile Program Cost Growth

Next, we examine missile programs in the same way in which we evaluated aircraft and space programs. As shown in Figure 5.3, all the missile programs are at or below the historical trend of completed programs' development cost growth.

Figure 5.3
Major Ongoing Missile Programs Compared to Trend of Completed Programs

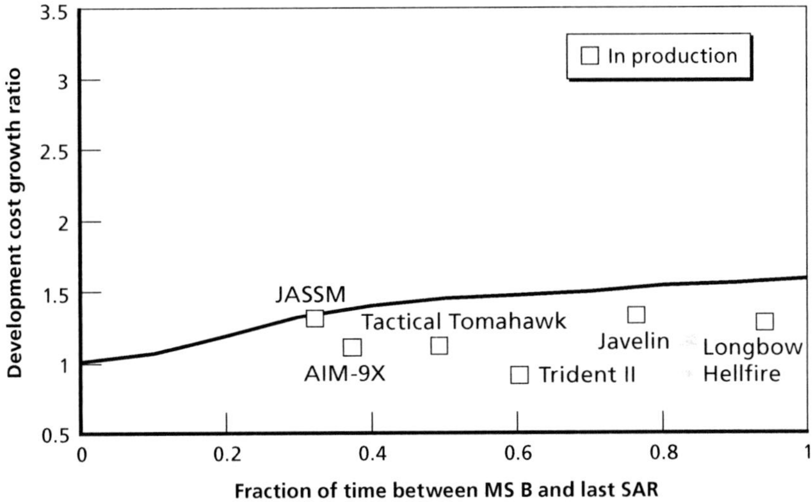

NOTE: JASSM = Joint Air-To-Surface Standoff Missile. AIM = air interceptor missile.
RAND *MG588-5.3*

Summary

The majority of ongoing aircraft and space programs have experienced higher-than-average cost growth in comparison to completed programs. On the other hand, missile programs are doing rather well and experience below-average development cost growth. Comparing the 19 ongoing programs shown in Figures 5.1 through 5.3 to historical averages indicates that eight are clearly doing worse, nine appear to be doing better, and two are in line with the averages. These results suggest that some recent, ongoing programs are experiencing higher growth than recent, completed ones. The differences lead us to conclude that, to examine the cost growth trend over the past three decades, ongoing programs must be incorporated into the analysis.

CHAPTER SIX

Is the Cost Growth Trend Increasing?

To gain a clearer picture of cost growth trends, we decided to include ongoing programs as well as completed ones. To ensure a fair basis for comparison, we examined ongoing and completed programs at similar points in their acquisition.

Additional Ongoing Programs

To include more recent programs in the analysis, we had to select a point at which programs were mature enough to include. Most cost growth occurs relatively early in a program's acquisition life. In fact, on average, 60 percent of cost growth occurred when the program was about one-third of the way between MS B and the last SAR. The approximate point at which this occurs is around five years past MS B; thus, this is the point at which we analyze both the completed and the ongoing programs.[1] The 33 ongoing programs with solid MS B estimates that were at least five years past their MS Bs as of the December 2004 SAR are as follows:[2]

---

[1] However, there are alternative methods to correct for maturity. For example, we compared cost growth at a common percentage completion in Arena, Leonard, et al. (2006). For this trend analysis, we could have examined growth at common percent completion, e.g., 30 percent of the total program duration past MS B. Our approach might slightly bias longer or larger programs toward lower growth (as they should be less mature at a fixed point). However, we observe no downward bias in the data.

[2] Appendix B provides a short synopsis of each program. As a result of the selection criteria (five years past MS B), the following programs were excluded from the analysis: JSF, Global

- aircraft
  - C-17
  - F/A-18E/F
  - F/A-22
  - JPATS
  - V-22
- electronics
  - Airborne Warning and Control System (AWACS) Radar System Improvement Program (RSIP)
  - B-1B computer
  - cooperative engagement capability
  - Global Broadcast Service (GBS)
  - JDAM
  - Joint Standoff Weapon (JSOW) baseline
  - JSOW unitary
  - maneuver control system
  - Multifunctional Information Distribution System–Low-Volume Terminal (MIDS-LVT)
  - Minuteman guidance replacement program
  - national airspace system computer
- helicopters
  - CH-47F improved cargo helicopter (ICH)
  - H-1 upgrades
  - Longbow Apache AF
  - MH-60R
  - MH-60S
- missiles
  - AIM-9X
  - Guided Multiple Launch Rocket System (GMLRS)
  - JASSM
  - Javelin
  - Longbow Hellfire
  - Patriot PAC-3 missile

Hawk, AEHF satellite, WGS, NPOESS, C-130 avionics modernization program, and C-5 RERP.

  - Tactical Tomahawk
  - Trident II missile
- space
  - EELV
  - Minuteman propulsion replacement program
  - SBIRS-Hi
- vehicles
  - family of medium tactical vehicles (FMTV).

Of the 33 programs, 11 are electronics, eight are missiles, five are aircraft, five are helicopters, three are space, and one is a vehicle. Once we include these ongoing programs, the sample size increases for the 1980s, though the mix remains similar; both the mix and sample size change substantially for the 1990s. Because all programs from the 1970s are complete, that decade's mix and sample size remain unchanged. Figure 6.1 shows the mix and quantities by decade of MS B, both with and without the ongoing programs.

Figure 6.1
Mix of Programs in the Completed Program Analysis Compared to Mix of Programs in the Completed Plus Ongoing Program Analysis

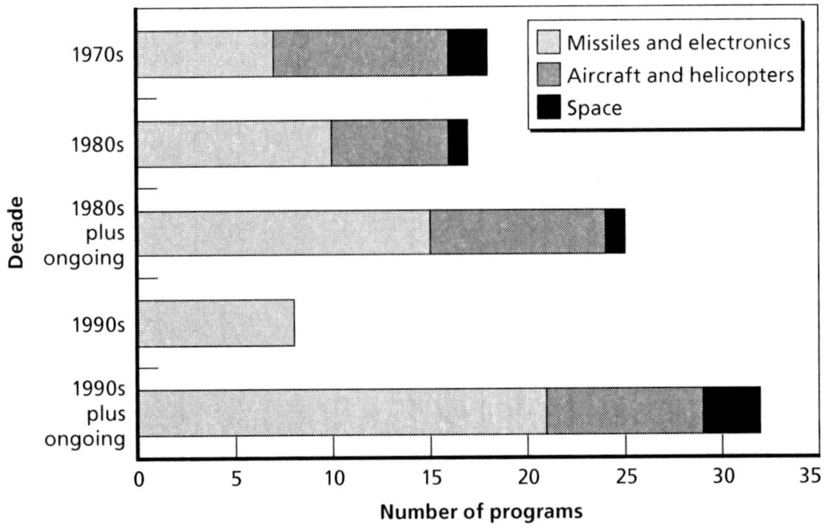

RAND *MG588-6.1*

Reexamining the DCGF Trend

Earlier DCGFs for each decade were calculated by dividing the program's final SAR total cost by the MS B estimate.[3] These figures were shown in Figure 4.3 in Chapter Four and are repeated for reference as the leftmost bars in each decade's grouping in Figure 6.2. Table 6.1 shows DCGF summary statistics at five years past MS B for both completed and ongoing programs by weapon type, and Tables 6.2 and 6.3 show the summary statistics by decade.

When adding the ongoing programs to the completed programs, the point of comparison for the calculation of the DCGF is five years past MS B. To facilitate this comparison, we recalculated the DCGF for the completed programs as a ratio of the estimate at that point

Table 6.1
DCGF Summary Statistics by Program Type

Program Type	n	Mean	Standard Deviation	Median
All programs	76	1.45	0.80	1.22
Aircraft	15	1.16	0.16	1.13
Cruise missiles	5	1.75	0.95	1.43
Electronic aircraft	5	1.59	0.31	1.65
Electronics	19	1.20	0.22	1.22
Helicopters	8	1.92	1.48	1.58
Launch vehicles	3	1.91	1.53	1.15
Missiles	14	1.50	1.04	1.30
Other	1	1.25	—	1.25
Satellites	3	1.64	0.50	1.88
Vehicles	3	1.81	1.06	1.21

[3] The final SAR total cost corresponds approximately to the 90-percent–complete point because SAR reporting is not required beyond that point.

Table 6.2
DCGF Summary Statistics by Decade

	Decade		
DCGF Statistic	**1970s[a]**	**1980s**	**1990s**
DCGF (n = 74)			
n	21	24	29
Mean (std. deviation)	1.49 (0.53)	1.37 (0.65)	1.24 (0.29)
DCGF (n = 76)			
n	21	24	31
Mean (std. deviation)	1.49 (0.53)	1.37 (0.65)	1.50 (1.04)

NOTE: n = 76 represents the entire sample. n = 74 represents the entire sample minus the two extreme observations (DCGF ≥ 5).

[a] Three programs from the 1960s are included with the 1970s in this analysis.

Table 6.3
Summary Statistics of DCGF Weighted by Development Dollars by Decade

	Decade		
DCGF Statistic	**1970s[a]**	**1980s**	**1990s**
DCGF (n = 74)			
n	21	24	29
Weighted mean (std. error)	1.36 (0.16)	1.30 (0.21)	1.25 (0.14)
DCGF (n = 76)			
n	21	24	31
Weighted mean (std. error)	1.36 (0.16)	1.30 (0.21)	1.30 (0.49)

NOTE: n = 76 represents the entire sample. n = 74 represents the entire sample minus the two extreme observations (DCGF ≥ 5).

[a] Three programs from the 1960s are included with the 1970s in this analysis.

to the MS B estimate, as opposed to what was discussed previously (in Chapters Three, in which the CGF was the ratio of estimate at the 90-percent–complete [final SAR] point to the MS B estimate for each of the decades). We selected the five-year point because most programs are beyond the early design stage and preliminary design review (PDR). Some programs, at five years beyond MS B, may have already passed their critical design review (CDR) as well.[4] Further, as indicated by the analysis of completed programs, most cost growth occurs early in the development phase. Thus, we deemed the five-year point to be the best point at which to measure the cost growth of both completed and ongoing programs. Table 6.2 presents DCGF summary statistics for each decade.

For each decade, we also calculated the mean DCGF weighted by the total development budget in constant dollars as reported in the SAR at about five years past MS B. These results are shown in Table 6.3.[5]

Adding the ongoing programs to the completed ones does not create a similar mix of programs across decades. To do this, we normalized these factors for the mix of systems included in each decade. We weighted the factors for the 1970s mix of systems and the 1990s mix of programs.

Figure 6.2 shows the mean value of the DCGF for the 1970s, 1980s, and 1990s. The first of bar in each decade shows the development cost growth of completed programs. Categorization into a particular decade is based on when a program had its MS B (or equivalent) decision. That set of bars indicates declining DCGF, from almost 1.8 in the 1970s to just over 1.2 in the 1990s. However, as discussed in Chapter Four, that trend should be regarded with a measure of skepticism,

---

4 PDR is a major system design milestone at which the government approves the weapon system's configuration and the contractor begins the design work. CDR is another design milestone at which most drawings of the approved configuration are about 80 percent complete and the contractor begins manufacturing the test hardware.

5 In Table 6.3, we report the weighted mean and the standard error; often, the variability around that mean is expressed in terms of standard deviations. Most of the population, especially if the normality assumption holds for this characteristic, lies within plus or minus two standard deviations of our mean. Such a mean and its associated standard deviation are called *descriptive statistics*. However, when we transition to weighted means, we enter the realm of "point estimates." The variability around point estimates is usually expressed as the standard error. Therefore, we present the standard error for the weighted mean.

Figure 6.2
Trend of Weapon System Development Cost Growth

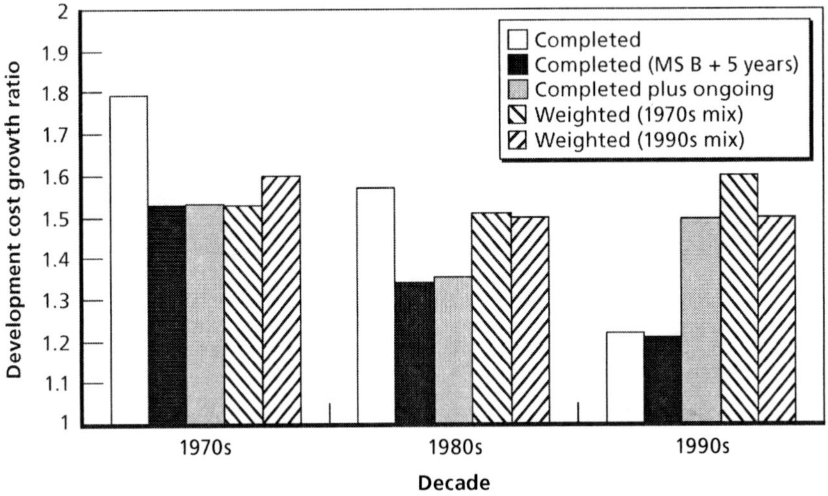

RAND *MG588-6.2*

in part because the averaged data for each decade include programs of different lengths (more recent programs are necessarily shorter) and in part because the mix of types of programs that are completed in each decade are quite different. For example, any project completed in the 1990s would have to be of shorter duration and would tend to be of the type that experiences the lowest cost growth (Arena, Leonard, et al., 2006). Previous research shows that shorter projects tend to have less cost growth than do longer ones. The research also shows that electronics programs tend to have less cost growth, and the vast majority of programs from the 1990s fall into that category. Thus, the apparent decline in cost growth may be influenced by the inclusion of shorter projects and weapon system classes that typically experience less cost growth.

The second bar for each decade grouping in Figure 6.2 traces the development cost growth trend by examining the same set of completed programs but measured at five years past MS B. As might be expected, measuring development cost growth from the five-year point does result in lower DCGFs. However, the relative ranking of the decades remains unchanged. Development cost growth still declines over time, but growth in the 1970s and 1980s is less because all growth after the five-year point is excluded and longer programs will not reflect all the growth in the program.

The third bar in each decade grouping incorporates 34 ongoing programs with the completed ones, all measured at five years past MS B. The combined sets of programs are compared at a common point: five years beyond MS B. These bars show that, although DCGF declined between the 1970s and the 1980s, from almost 1.49 to 1.37, in the 1990s, DCGF rose again to about the same level as in the 1970s—1.50. In addition, the large increase in development cost growth from the second to the third bar in the 1990s grouping indicates that ongoing programs begun in the 1990s have substantially higher cost growth than those begun in that decade that are now complete.

Our research also shows that different program types have different cost growth values. Previous research (Arena, Leonard, et al. 2006; McNicol, 2004; Drezner et al., 1993) has indicated significant differences among types of weapon systems (e.g., aircraft, missiles, electronics). To assess the effect of weapon program type, we controlled for type by normalizing the contribution of each program type to DCGF based on the proportions in the 1970s or the 1990s. There were four steps to this normalization process. First, we grouped projects by similar weapon system type. The groupings were (1) all aircraft and helicopters; (2) launch vehicles and satellites; and (3) missiles, electronics, and other programs. We determined these groups based only on our assumptions as to similarity of type. Ideally, we would have liked not to resort to groupings. However, each weapon system type was not represented in each decade, so some aggregation was required.

With the three functional groupings in hand, we determined both the frequency counts and simple average growth ratio for the three groupings for each decade—the second and third steps. The weightings were based on the percentage of projects making up that category for the decade that served as the normalization baseline. In the 1970s, the weights were 50 percent for group one (i.e., nine out of 18 projects), 11 percent for group two, and 39 percent for group three. The last step was to multiply the average growth for the group by the normalization weight and sum. For example, the 1970s weighted growth for the 1980s projects is $1.19 \times 0.5 + 3.67 \times 0.11 + 1.31 \times 0.39 = 1.51$. This process was repeated for each decade.

These normalized cost growth figures are shown in the fourth and fifth bars in each decade's grouping in Figure 6.2. The fourth bar shows completed and ongoing programs all measured at five years past MS B and weighted by the 1970s program mix. Similarly, the fifth bar shows completed and ongoing programs all measured at five years past MS B and weighted by the 1990s program mix. These two sets of bars show that development cost growth has not improved over the decades, with the possibility of a slight decrease in the 1980s followed by a slight increase in the 1990s. Controlling for the 1970s mix of programs shows some increase between the 1970s and the 1990s, but not a significant one. Controlling for the 1990s program mix shows a slight decline between the 1970s and the 1990s. Because of limitations of our data, we grouped programs by similar functionality (e.g., aircraft and helicopters, satellites and launch vehicles). There are many other combinations of system types (e.g., by service, cost), but we explored only one such grouping in this research. However, in Appendix C, we do a formal statistical analysis in which we examine trends based on the groups and on system type. The system type, by and large, does not show any significant time trend for any of the weapon system types other than helicopters (for the nonweighted analysis) and satellites and launch vehicles (for the weighted analysis). This grouping is the most detailed level of grouping—we would not expect any higher-level grouping to show a trend, either.[6] The analysis did not change our basic conclusion that the cost growth in the past three decades has been high and that there has been no improvement over the past three decades.

Figure 6.3 represents a box plot of the DCGF by decade, which, without any adjustments for program mix, would lead us to the same

6 The statistically significant trends for helicopters and space programs are driven by single observations. For the detailed statistical analysis, see Appendixes C and D. The statistical analysis uses both DCGF and DCGF weighted by the development budget in constant dollars reported in the SAR five years past MS B.

conclusion. However, as suggested by Figure 6.3, variability[7] in DCGF is greater in the 1990s than in the earlier decades.[8]

We also conducted a statistical analysis (see Appendix D) to examine whether there were any differences in the trend of development cost growth among the military services, and we conclude that the three services do not differ significantly in their DGCF levels for their respective programs.[9]

Figure 6.3
DCGF Box Plot by Decade

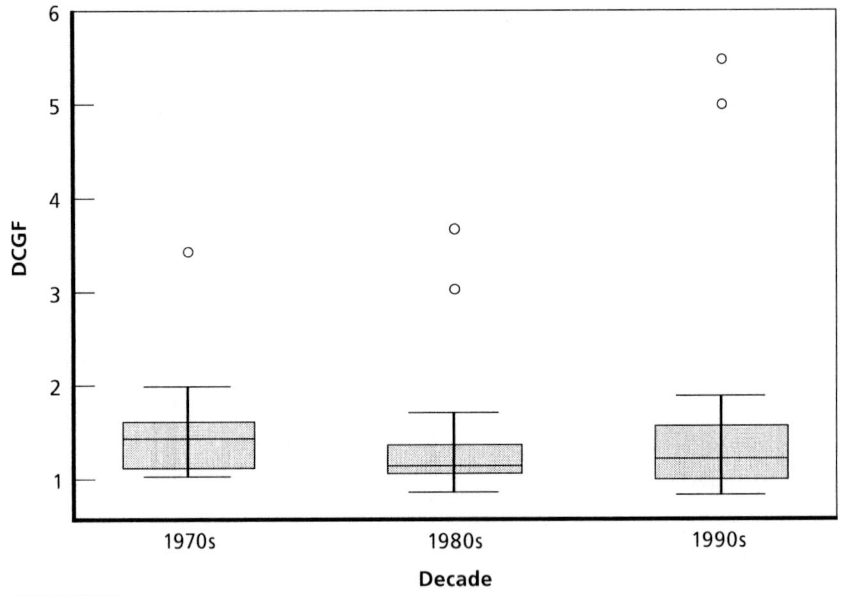

RAND *MG588-6.3*

---

[7] *Variability* refers to the degree to which a collection of observations is dispersed around some central tendency.

[8] Appendix C includes the statistical analysis.

[9] This analysis also included both the DCGF and the DCGF weighted by the development budget in constant dollars reported in the SAR submitted around five years past MS B.

CHAPTER SEVEN

How Does Defense System Cost Growth Compare to That of Other Systems?

As mentioned previously, weapon system cost growth has been the topic of acquisition-related research for many years. However, this phenomenon is not unique to military systems: It also appears in the development of other complex systems. This chapter provides a summary of other megaprojects and their cost growth. The cost growth figures are not quite comparable because some studies did not control for differences in inflation or the mix of systems produced in each decade.

Public Works Projects

Unfortunately, cost information for civilian projects is more difficult to obtain than it is for weapon systems. Cost information is typically considered highly proprietary and competition-sensitive for commercial projects such as civil aircraft development and the automotive industry. For this comparison, we had to rely on data that were readily available for other federally, municipally, or state-run projects.

A recent study of the cost growth of 258 transportation projects (Flyvbjerg, Holm, and Buhl, 2002), which included rail, fixed-link, and road projects,[1] reported an average of 27.6-percent cost growth covering a period of 70 years. Table 7.1 summarizes the statistical analyses of the study.

---

[1] *Rail* includes urban high-speed and intercity conventional high-speed rail systems; *fixed-link* includes bridges and tunnels; *roads* include highways and freeways.

Table 7.1
Cost Growth of Transportation Projects

Project Type	n	Average Cost Growth (%)	Standard Deviation	Level of Significance
Rail	58	44.7	38.4	<0.001
Fixed-link	33	33.8	62.4	<0.004
Road	167	20.4	29.9	<0.001
All projects	258	27.6	38.7	<0.001

SOURCE: Flyvbjerg, Holm, and Buhl (2002).

That study also found that the vast majority—nine out of 10—of the transportation projects had experienced cost growth and that cost underestimation has not decreased with time. Interestingly, both these findings are similar to our weapon system cost growth findings.

The city of Boston's central artery/tunnel project (known as the "Big Dig") and the commonwealth of Virginia's Springfield Interchange (known as the "Springfield Mixing Bowl") are not quite complete;[2] however, both projects have experienced considerable cost growth in comparison to their original estimates. Other famous public works projects, such as the Washington, D.C., Metro and the Boston–Washington–New York rail system, also experienced substantial cost growth (Flyvbjerg, Bruzelius, and Rothengatter, 2003). In comparison to all these projects, the development cost growth of weapon systems is lower. Table 7.2 lists the CGF for each of these projects.

Civilian Space Projects

A 1992 U.S. General Accounting Office study evaluated 29 National Aeronautics and Space Administration (NASA) programs with more than $200 million in development costs. These programs were

---

2 For information about the Big Dig, see Massachusetts Turnpike Authority (undated); for information about the Springfield Mixing Bowl, see Virginia Department of Transportation (undated).

Table 7.2
Cost Growth of Some Recent Public Works Projects

Project	CGF
Big Dig	2.77
Springfield Interchange	2.40
Washington, D.C., Metro	1.85
Boston–Washington–New York rail	2.30

SOURCE: Flyvbjerg, Bruzelius, and Rothengatter (2003); Massachussetts Turnpike Authority (undated); Virginia Department of Transportation (undated).

initiated between 1977 and 1991 (U.S. General Accounting Office, 1992). The analysis included space projects, such as planetary missions, space and earth science missions, manned missions and related programs, and other programs such as communication satellites. That study reported cost growth experience for 25 of the 29 projects when adjusting the initial cost estimates and the final cost estimates using NASA's inflation indexes. The median growth was reported to be about 77 percent. A more recent U.S. General Accounting Office study examined 27 development space projects and reported that more than half of these projects experienced cost increases, some as much as 94 percent (U.S. General Accounting Office, 2004).

Gas and Oil Projects

A recent evaluation of cost and schedule performance of the upstream gas and oil industry suggests a much more modest cost increase in comparison to the weapon systems, public works projects, and the civilian NASA programs. The average industry cost growth is reported as 10 percent where a 30 percent cost growth is considered a disaster (Cotrill, 2003).

In a 1981 RAND study, Merrow, Phillips, and Myers (1981) explored the cost growth and performance shortfalls of 44 pioneer process plants in the oil and chemical industries. The study found that most of the cost growth of these projects was related to the extent to which the technology departed from what was currently proven, the degree of definition in the plant's site and related characteristics, and the complexity of the plant. On average, the cost growth for these 44 plants was 28 percent above that estimated (with a range of 9 to 59 percent, plus or minus one standard deviation) after correcting for inflation and scope (discretionary) changes. Similarly, Merrow, McDonnell, and Arguden (1988) found that, for 47 very large civilian projects (>1 billion 1984 dollars), the average cost growth was 88 percent. They attributed this growth to regulatory issues, the degree of technical innovation, type of ownership (e.g., public, private, or joint), and the extent of existing infrastructure.

Summary

Cost growth is not unique to defense weapon systems; rather, it is an inherent problem in many complex projects. Furthermore, a very rough comparison with the extensive analysis by Flyvbjerg, Holm, and Buhl (2002) of public works projects leads us to conclude that weapon system cost growth is higher than that of rail, fixed link, and road projects. Although there are examples of large, technically complex civilian projects with greater cost growth than that of the average DoD program (e.g., Boston's Big Dig; the Springfield Interchange; and the Washington, D.C., Metro), one should avoid drawing conclusions based on this comparison because some individual DoD programs exceeded the growth experienced in these projects.[3] Also, weapon system development cost growth has been relatively comparable to that of NASA space programs but higher than that of offshore gas and oil industry projects.

---

3 The DCGF range of weapon systems included in our analysis of completed weapon systems was 0.77 to 5.47 (Arena, Leonard, et al., 2006).

CHAPTER EIGHT

Conclusions

Despite many efforts and numerous recommendations to improve the acquisition process, cost growth of DoD weapon systems remains high. Development cost growth for completed programs is about 60 percent, most of which occurs early in the development phase. Further, our analysis of completed programs shows that longer programs experience higher cost growth than the average DCGF for all programs. Moreover, electronics programs have the lowest cost growth, and that difference was statistically significant.

Perhaps the most important finding of this analysis is that the trend of the average development cost growth for all weapon systems included in our data over the past three decades has remained high, without any significant improvement. Further, when we weight the DCGF by the SAR-stated development budget in constant dollars at the five-year point, our observation on the overall trend does not change; however, our statistical analysis in Appendix C indicates that launch vehicles (three observations) and satellites (three observations) do show a statistically significant higher trend. Although the results are statistically significant, the small number of observations included in the analyses means that extreme caution should be used in making any generalizations about these particular results. Thus, the many acquisition reform initiatives have not succeeded in lowering the development cost growth of military systems. This is not an indictment of the government personnel or contractors involved in the acquisition of systems for the military, but, rather, is related to the overoptimism of program managers, the complexity of the technology, the scale of the program, and other factors.

As our very rough comparison with the analysis by Flyvbjerg, Holm, and Buhl (2002) of public works projects shows, weapon system total cost growth is higher than that of rail, fixed-link, and road projects. The difference is understandable given that the technologies involved in those projects are extremely well understood and include the conservative, evolutionary adaptation of new technology over time. Most DoD defense development programs involve much higher levels of new technology adaptation and therefore result in inherently higher levels of cost and schedule uncertainty.

This is also not to say that DoD cannot do better in controlling cost growth. Undoubtedly it can, and it should bend its efforts to doing so. Over the years, several studies, by RAND and others,[1] have attempted to identify the causes of cost growth and what steps can be taken to address them. These causes fall into the following broad areas: overoptimism, estimating errors, unrecognized technical issues, requirements creep, lack of incentives to control cost, and schedule extensions. Therefore, addressing the issue of cost growth requires vigorous involvement of all stakeholders in DoD. However, there may be one aspect of the acquisition process that merits special attention, and that is the cost estimates that form the basis of program budgets. Cost growth may also reflect poor initial budget estimates. Better cost estimates would not necessarily save any money; however, they would provide decisionmakers a better basis for deciding whether to pursue a given program.

Future Research

Data collection and analysis for this study were concluded in 2005, so any future analysis of cost growth should include the latest SAR data

[1] Recent RAND studies of ways in which DoD can address the growing cost of weapon system development include the following: Arena, Younossi, et al. (2006); Arena, Blickstein, et al. (2006); Younossi, Stem, et al. (2005); Lorell and Graser (2001); and Cook and Graser (2001).

now available; this will allow for programs such JSF, Global Hawk, AEHF satellite, WGS, NPOESS, the C-130 avionics modernization program, and the C-5 RERP to be included in the analysis. These programs were excluded because, at the time of this research, they had not reached the five-year point beyond MS B. Further, our choice of five-year point past MS B was based on our analysis that showed that most cost growth occurs early in a weapon system's development, but, if new data allow for the addition of more programs past that point, sensitivity analysis of the point at five years past MS B may be prudent.

APPENDIX A

Completed Programs

Aircraft

A-7D Corsair II

Data are from September 1969 through June 1975 "A-7D" SARs. The MS III baseline is solid.

A-10 Specialized Close Air Support Aircraft

Data are from March 1973 through March 1982 "A-10" SARs. The September 1971 through September 1972 "A-X (A-9/10)" SARs report prototype phase costs only; thus, no program baseline estimates are available from the MS I equivalent. The first "A-10" SAR is dated December 1972 and is not used in the cost growth track as it is pre–MS II. The program has solid MS II and MS III baselines.

AV-8B Harrier II

Data are from the June 1981 through December 1992 "AV-8B" and "AV-8B Harrier II" SARs. This program should not be considered a joint program, as the development of the UK Royal Air Force's GR 5 did not begin until August 1981, four months after production contract awards for the U.S. Navy's AV-8B. The MS III baseline is solid.

AV-8B Harrier Remanufacture

Data are from the December 1994 through December 2002 "AV-8B Remanufacture" SARs. The MS III is concurrent with the initial SAR report and the baseline is solid.

B-1B Lancer

Data are from the December 1981 through December 1992 "B-1B" SARs. MILCON costs are not included in the database, as they are reported in most of the SARs in a footnote and only in TY dollars. A single contract was awarded for the B-1B's development and production. The MS II/III baseline is solid.

B-2A Spirit

Data are from the December 1996 SAR. The December 1996 SAR indicates that prior SARs were generated, but RAND does not have access to them. The MS III baseline is solid. However, no baseline CIC can be determined and there are only two meaningful estimates: the baseline and the final. Due to the large quantity reduction—from 127 to 15—the normalized CGFs (procurement and program) are not suitable for database inclusion and analysis. In addition, six R&D aircraft were later adapted to production configuration.

C-5B Galaxy

Data are from the June 1983 through December 1988 "C-5B" SARs. This program is a follow-on procurement of the C-5A program. The MS III baseline is solid.

E-2C Hawkeye/Carrier-Based Airborne Early Warning Command and Control System

Data are from the September 1971 through December 1989 "E-2C" and December 1990 through December 1991 "E-2C AEW (Hawkeye)" SARs. Reports were suspended between June 1981 and December 1984. TY dollar estimates are erroneously reported as BY dollars in the December 1984 through December 1990 SARs. These errors are corrected in the SAR cost growth track using data from the June 1981 and December 1991 SARs. BY program cost figures are estimates made prior to the September 1975 SAR. The MS III baseline is solid.

AWACS E-3A Sentry

Data are from the March 1970 through September 1970 "411 L Airborne Warning and Control System" and December 1970 through June

1984 "E-3A Airborne Warning and Control System" SARs. Data from the September 1969 and December 1969 SARs are not useful. A full program estimate is given at MS I, but only the DCGF is calculated because the validity of procurement cost estimates (and their BY dollar values) cannot be established in the earliest SARs. Regarding MS II, annual data of sufficient quality to establish a CIC were not available until the December 1974 SAR. Thus, only unadjusted CGFs are available for those observations prior to December 1974, and at no point can adjusted procurement CGFs using the baseline CIC be calculated, as there is no baseline CIC. All MS III–related data are solid; thus, a full set of CICs is calculated. All three milestones have solid baselines.

Advanced Airborne Command Post (AABNCP) National Emergency Airborne Command Post (NEACP) E-4

Data are from the March 1973 through March 1982 "E-4" SARs. This program procured new Boeing 747 aircraft and outfitted them with electronics, creating an airborne command-and-control system. The program's MS II/III baseline is solid.

E-6A Airborne Strategic Communication Aircraft

Data are from the June 1983 through December 1983 "E-6," December 1984 through December 1989 "E-6A," and December 1990 through December 1991 "E-6A TACOMO" SARs. The program's MS II and MS III baselines are solid.

E-8A Joint Surveillance Target Attack Radar System (JSTARS)—Airborne Radar Segment

Data are from the December 1985 through December 2003 "Joint STARS" SARs. The ground segment is Army-funded and is tracked separately. The MS II and MS III baselines are solid.

EF-111A Tactical Jamming System (TJS)

Data are from the March 1976 through March 1979 "EF-111A" and June 1979 through December 1983 "EF-111A TJS" SARs. The program has solid MS II and MS III baselines.

F-14A and F-14A Plus

Data are from the June 1969 through September 1983 "F-14" and "F-14A," December 1983 through December 1985 "F-14A/D," and December 1986 "F-14A" SARs. There was a break in reporting between September 1972 and June 1973. The F-14D and F-14 Block 1 are tracked separately. The MS II baseline is solid and the adjustments for quantity changes are estimated reasonably well. The MS III baseline and associated adjustments for quantity changes are estimated reasonably well. Both the MS II and MS III SAR cost tracks are estimated well enough to be suitable for database inclusion, and analysis and the baselines are solid.

F-14D Tomcat

This program was formerly known as the F-14A Plus. Data are from the December 1986 through December 1993 "F-14D" SARs; data from the December 1983 through December 1985 "F-14A/D" SARs are omitted because the SARs do not break out F-14D costs from those of the F-14A. The program began by planning to build only new aircraft, but it then changed to a mix of new builds and the remanufacture of existing aircraft. The MS II baseline was based on new aircraft only and was specified two years after the initiation of the program. The program average unit cost of the new aircraft is 150 percent that of the mix of new and used aircraft that are represented in all other observations. This makes it impossible to adjust for the quantity change that occurred in the second SAR, and it makes the MS II baseline unsuitable for inclusion in the database and analysis. The MS III baseline is solid.

F-15A/B/C/S/E Eagle

Data are from the December 1969 through December 1990 "F-15," "F-15 Advanced Tactical Fighter (Air Superiority)," "F-15 Advanced Tactical Fighter," and "F-15 (Eagle)" SARs. Both the MS II and MS III baselines are solid.

F-16 Falcon

Data are from the December 1975 through December 1994 "F-16" SARs. At four distinct points in the production run, a follow-on fighter was funded within the F-16 program and then quickly canceled. The distortion of these funding additions and subsequent subtractions has been eliminated from the SAR track by excluding them from the program totals. Both the MS II and MS III baselines are solid.

F-5E Fighter Aircraft

Data are from the December 1971 through March 1976 "F-5E" SARs. An MS II could not be established because SARs prior to December 1971 contained aggregate costs and quantities for Air Force and foreign military sales customers. The MS III baseline is solid.

F/A-18A/B/C/D Fighter Aircraft

Data are from the March 1976 through September 1979 "F-18," December 1979 through March 1991 "F/A-18" and "Hornet (F/A-18)," and December 1991 through December 1994 "F/A-18 C/D" SARs. The MS II and MS III baselines are solid.

KC-135R Reengine

Also known as the strategic tanker modification program, the program purchases and installs new engines and strengthens main landing gear and makes other system improvements to the tanker aircraft. Data are from the December 1982 through December 1989 "KC-135R" and December 1990 through December 1994 "KC-135R Reengine" SARs. The MS III baseline is solid.

S-3A Carrier-Based Antisubmarine Warfare Aircraft

Data are from the June 1969 through March 1977 "S-3A" SARs. The MS II and MS III baselines are solid.

T-45 (GOSHAWK) Undergraduate Jet Flight Training System (T45TS)

Data are from the December 1983 and March 1984 "T45TS Undergraduate Jet Flight Training System," June 1984 and September 1984 "T45TS (formerly VTXTS) Undergraduate Jet Flight Training System," December 1984 "Navy Undergraduate Jet Flight Training System (T45TS)," and December 1985 through December 2004 "T45TS" SARs. This program's cost growth is significantly affected by exchange rate fluctuations. The MS I baseline was established 16 months after the MS I contract award and just nine months before the MS II contract award. It is not appropriate for database inclusion and analysis. The MS II and MS III baselines are solid.

Helicopters

CH-53E Super Stallion; MH-53E Sea Dragon

Data are from the June 1973 through June 1984 "CH-53E" and December 1984 through December 1994 "C/MH-53E" SARs. The two helicopters are very similar and are reported in aggregate. The MS I, MS II, and MS III baselines are solid.

OH-58D Kiowa Warrior Army Helicopter Improvement Program (AHIP)

Data are from the September 1982, December 1982, and June 1989 "Army Helicopter Improvement Program"; December 1983 through December 1988 "Army Helicopter Improvement Program (AHIP)"; December 1989 "Army Helicopter Improvement Program (AHIP) (Kiowa Warrior)"; December 1990 and December 1991 "AHIP Kiowa Warrior"; and December 1992 through June 1995 "OH-58D Kiowa Warrior" SARs. The MS II and MS III baselines are solid.

UH-60A/L "Black Hawk" (Utility Tactical Transport Aircraft System [UTTAS])

Data are from the September 1972 through September 1976 "UTTAS," December 1976 through June 1977 "UH-60A (UTTAS)," September 1977 through March 1979 "UH-60A (Black Hawk)," June 1979 through September 1982 "Black Hawk," December 1982 and December 1983 "UH-60A," December 1984 through December 1988 "UH-60A Black Hawk," December 1989 "UH-60A/UH-60L Black Hawk,"

December 1990 through December 1994 "UH-60A/L Black Hawk," and December 1995 through December 1999 "UH-60L Black Hawk" SARs. The MS II and MS III are solid.

Launch Vehicles

Inertial Upper-Stage (IUS) Rocket

Data are from the December 1982 through December 1990 "IUS," December 1991 "Space Shuttle (IUS)," and December 1992 and December 1993 "Inertial Upper Stage" SARs. The IUS is a two-stage, solid-propellant rocket vehicle used to deploy satellites to higher orbits from the space shuttle payload bay or atop the Titan IV booster. SARs report that quantities and funding are for Air Force units only, but the IUS was also procured for NASA and an unspecified DoD agency. The SAR states a single development unit, but at least seven were built and flown for the three customers. The MS III baseline is solid.

Peacekeeper (LGM-118)

Data are from the December 1983 through December 1989 "Peacekeeper" and December 1990 through December 1992 "PCKR in MM Silo" SARs. Development and MILCON costs prior to FY 1983 and FY 1984, respectively, are reported only in TY dollars. These costs are approximated in BY dollars and added to the program cost track. This program includes missiles and missile initial spares for rail garrison basing mode, but the trains and associated shelters for that mode are reported in the Peacekeeper rail garrison SARs. U.S. Department of Energy nuclear warhead costs are excluded from this SAR track. The MS III baseline is solid.

Minuteman III Intercontinental Ballistic Missile (ICBM) (WS-133B) (LGM 30G)

Data are from the June 1969 "Minuteman III," September 1969 and December 1969 "WS-133B LGM-30G," and March 1970 through March 1978 "LGM 30G" SARs. There is no precise date for the MS

III, but the date of the SAR baseline cost estimate has been determined to be within a few months of the MS III equivalent. The MS III baseline is solid.

Titan IV Complementary Expendable Launch Vehicle (CELV)

Data are from the December 1985 "CELV," September 1986, December 1986 and December 1990 through December 1991 "Titan IV (CELV)," September 1992 and December 1992 "Titan (ELV)," and December 1987 through December 1989 and December 1993 through December 2001 "Titan IV" SARs. A single contract for both RDT&E and production was awarded on February 28, 1985. The date of the baseline estimate in the initial SAR is one year after the MS II and MS III contract award. This baseline date is close enough to the contract award date for the baseline to be considered solid; thus, the MS II/III is suitable for database inclusion and analysis.

Missiles

ACM Advanced Cruise Missile (AGM-129A)

Data are from the December 1989 through December 1992 "AGM-129A/Advanced Cruise Missile (ACM)" and December 1990 through December 1992 "ACM (AGM-129)" SARs. The December 1989 SAR indicates that earlier SARs exist, but these SARs are not available. The MS III contract was awarded on November 1, 1984, more than five years prior to the first available SAR. The baseline in the earliest available SAR is dated December 1987, more than two years after the MS III contract award. The program is omitted from the data set because no baseline in the proper time frame is available.

AGM-65A/B Maverick

Data are from the March 1969 through September 1974 "AGM-65A" and December 1974 through September 1976 "AGM-65A/B" SARs. This Maverick is a TV-guided, rocket-propelled, air-to-ground missile. This program contains solid MS II and MS III baselines.

AGM-65D Maverick

Data are from the December 1978 through March 1980 "Imaging Infrared (IIR) Maverick," June 1980 through September 1985 "Infrared (IR) Maverick," December 1985 through December 1989 "AGM-65D & AGM-65G," and December 1990 through December 1992 "IR Maverick" SARs. Data from the December 1976 through September 1978 IIR Maverick SARs are omitted, as these reports were prior to MS II. The program contains solid MS II and MS III baselines.

AGM-88A/B/C High-Speed Antiradiation Missile (HARM)

Early program data are from the Navy's September 1978 through December 1986 "HARM (AGM-88A)" and the Air Force's December 1978 through December 1983 "AGM-88" and December 1984 through December 1986 "HARM (AGM-88A)" SARs. Data for both services are from the December 1987 "HARM (AGM-88A/B)" and December 1988 through June 1994 "HARM (AGM-88A/B/C)" SARs. The MS II and MS III baselines are solid.

AIM-54C Phoenix

Phoenix C is an improved version of the Phoenix A, which is a long-range tactical air-to-air missile carried only by the F-14 fighter aircraft. Data are from the September 1977 and December 1977 "Phoenix," March 1978 through March 1981 "AIM-54A/C Phoenix," and June 1981 through December 1991 "Phoenix (AIM-54C)" SARs. Funding for retrofit of the AIM-54A missile is not included in this SAR track. Data from the June 1976 through June 1977 "Phoenix" SARs are omitted because they are pre–MS II. The program contains solid MS II and MS III baselines.

Air-Launched Cruise Missile ALCM (AGM-86)

Data are from the December 1977 through March 1979 "Air Launched Cruise Missile," June 1979 through June 1983 "Air Launched Cruise Missile (ALCM)," December 1983 through December 1984 "ALCM," and December 1985 "(Air Launched Cruise Missile)" SARs. The December 1982 SAR indicates that it is the last, but program SARs were reinstated in June 1983. The costs of nuclear components within the system are omitted. The program contains solid MS II and MS III baselines.

Advanced Medium-Range Air-to-Air Missile (AMRAAM) (AIM-120)

Data are from December 1982 through December 2004 "AMRAAM" and "AMRAAM (AIM-120)" SARs. The program contains solid MS II and MS III baselines.

Ground-Launched Cruise Missile (GLCM) (BGM-109G)

The system is derived from the BGM-109B Tomahawk. Data are from the December 1977 through March 1979 "Ground Launched Cruise Missile," June 1979 through March 1983 "Ground Launched Cruise Missile (GLCM)," December 1983 "GLCM," and December 1984 through December 1988 "Ground Launched Cruise Missile (BGM-109G)" SARs. The SAR costs cover the new-build Tomahawk (modified design) missiles, the transporter-ejector-launcher, and the launch control center. The costs of nuclear components within the system are omitted. The MS II and MS III baselines are solid.

Harpoon (AGM-84) Air-Launched, (RGM-84) Ship-Launched, (UGM-84) Submarine-Launched, and Standoff Land Attack Missile

Data are from the September 1971 through March 1979 "Harpoon," June 1979 through March 1982 "AGM-84A/RGM-84A/UGM-85A Harpoon," June 1982 through June 1990 "Harpoon (AGM/RGM/UGM-84A/C/D)," and December 1990 through December 1991 "Harpoon (A/R/UGM-84)" SARs. The MS II and MS III baseline dates were chosen for the first contract that expended significant developmental and production dollars, respectively. The milestones are solid.

Javelin Advanced Antitank Weapon System–Medium (AAWS-M)

Data are from the September 1989 through December 1989 "Advanced Antitank Weapons System–Medium," December 1990 "AAWS-M (Medium)," September 1991 through December 1995 "Javelin (AAWS-

M)," and December 1996 and subsequent "Javelin" SARs. The MS II and MS III baselines are solid.

Longbow Hellfire Antitank Air-to-Ground Missile

This is a subsystem of the AH-64 Apache Weapon System. Data are from the December 1990 through December 2004 "Longbow Hellfire" SARs. The MS II and MS III baselines are solid.

Army Tactical Missile System (ATACMS–Block I–APAM) (MGM-140A)

Data are from the September 1986 through December 1988 "Army Tactical Missile System (Army TACMS)," December 1989 through December 1993 "Army TACMS," and December 1994 through December 1999 "Army TACMS/APAM" SARs. Data from the September 1984 and December 1984 "Joint Tactical Missile System (JTACMS)" and the December 1985 "Army Tactical Missile System" SARs are omitted, as these reports were pre–MS II. The MS II and MS III baselines are solid.

Tomahawk Sea-Launched Cruise Missile (B/R/UGM-109)

Data are from the December 1977 through March 1979, December 1982 through June 1983, and March 1984 through December 1984 "Tomahawk Sea Launched Cruise Missile, BGM-109"; June 1979 through September 1982 "Tomahawk Sea Launched Cruise Missile"; December 1983 "Tomahawk (BGM-109)"; December 1985 through December 1989 "Tomahawk Sea Launched Cruise Missile, R/UGM-109"; and December 1990 through December 1996 "Tomahawk (R/UGM-109)" SARs. The MS II is designated by a Defense Systems Acquisition Review Council decision memorandum that was issued roughly halfway through the period in which the five developmental contracts were let; the MS III was determined by the contract award date to the prime manufacturer of the missile's airframe. The MS II's suitability for database inclusion is questionable because of the time lapse between the initial developmental contract and the first available program estimate. The MS III is solid.

Electronics

Low-Altitude Navigation and Targeting Infrared System for Night (LANTIRN) (AAQ-11/12)

Data are from the December 1982 through December 1993 "LANTIRN" SARs. The MS II baseline estimate was established after the MS II–equivalent date; thus, it may not be suitable for database inclusion and analysis. The MS III baseline is solid.

Advanced Field Artillery Tactical Data System (AFATDS)

Data are from the December 1990 through December 1998 "AFATDS" SARs. The MS II and MS III baselines are solid.

B-1B Conventional Mission Upgrade Program—Computer Upgrade

Data are from the December 1996 and December 1997 "B-1 CMUP-Computer" and December 1998 through December 2004 "B-1B CMUP" SARs. The program replaces six existing onboard computers with four new ones. The MS II and MS III baselines are solid.

B-1B Conventional Mission Upgrade Program—JDAM/1760/GPS/Communications

Data are from the December 1996 and December 1997 "B-1 CMUP-JDAM" and December 1998 through December 1999 "B-1B CMUP" SARs. The program integrates aircraft JDAM delivery capability with associated GPS and communication improvements. The MS II and MS III baselines are solid.

E-3 Sentry AWACS RSIP

Data are from the December 1989 "Radar System Improvement Program (RSIP)," December 1990 through December 1992 "AWACS RSIP," and December 1993 and subsequent "E-3 AWACS RSIP" SARs. The program updates the aircraft's radar hardware and software. The MS II and MS III baselines are solid.

Forward Antiaircraft Defense System—Command, Control, and Intelligence

This program's prior name was SHORAD C2—short-range air defense command and control systems. Data are from the September 1984 through December 1985 "Short Range Air Defense Command and Control System (SHORAD C2)," December 1986 and December 1987 "Forward Area Air Defense System (FAADS) Forward Area Air Defense Command, Control, and Intelligence (FAAD C2I)," December 1988 through December 1989 "Forward Area Air Defense Command, Control, and Intelligence (FAAD C2I)," and December 1990 through December 1998 "FAAD C2I" SARs. In the first third of the program, no quantity data were specified; in the remainder, the unit quantity definition fundamentally changed and then evolved. As a result, no quantity-normalized CGFs can be generated. All three MS baselines are solid. SAR reporting ceased because the program was broken into two pieces. The early portion was 100-percent delivered (thus no longer requiring reporting), and the later portion was not large enough to meet the SAR dollar thresholds.

Forward-Area Air Defense System—Line of Sight–Rear (LOS-R)

This system was known as the pedestal-mounted Stinger early in the program and Avenger later in the program. Data are from the December 1987 through December 1988 "Forward Area Air Defense System (FAADS) Line of Sight-Rear (LOS-R)," December 1989 and June 1990 "Forward Area Air Defense System (FAADS) AVENGER (LOS-R)," and December 1990 through September 1995 "AVENGER" SARs. The MS III baseline is solid.

Joint Service Imagery Processing System (JSIPS)

Data are from the December 1993 through December 1997 "JSIPS" and December 1998 "JSIPS (CIGSS)" SARs. The MS III baseline is solid and appropriate for database inclusion and analysis.

JSTARS—Common Ground Station (CGS)

Data are from the December 1988 through December 1990 "Joint STARS" Air Force SARs and the December 1991 "Army Joint STARS

GSM," December 1992 through December 1997 "Joint STARS GSM," December 1998 through September 2000 "Joint STARS CGS," and December 2001 "CGS" Army SARs. The airborne segment (E-8A) is Air Force–funded and is tracked separately. The CGS is the follow-on ground station—the original was the GSM, or ground station module, which is also tracked separately. All three MS baselines are solid and suitable for database inclusion and analysis.

JSTARS—Ground Station Module (GSM)

Data are from the December 1984 "Joint Surveillance and Target Attack Radar System (Joint STARS)," December 1985 through December 1990 "Joint STARS," December 1991 "Army Joint STARS GSM," December 1992 through December 1997 "Joint STARS GSM," and December 1998 through September 2000 "Joint STARS CGS" SARs. The December 1984 through December 1990 SARs are Air Force; the December 1991 through September 2000 SARs are Army. Data in the December 1996 and subsequent SARs are not relevant, as they cover the follow-on ground station, the CGS, which is tracked separately. The airborne segment (E-8A) is Air Force–funded and is tracked separately. The MS II baseline for development was established after the MS II contract award; thus, it might not be suitable for database inclusion and analysis. There was no procurement estimate until shortly before the MS III equivalent. The MS III baseline is solid.

Joint Tactical Information Distribution System (JTIDS), Class 2

The JTIDS program comprises the time division multiple-access (TDMA) terminal (Air Force/Army/U.S. Marine Corps [USMC]) and distributed time division multiple-access (DTDMA) terminal (Navy). The program had separate service SARs during its first five years and service-integrated SARs in its remaining years. The data are taken from the June 1982 through December 1983 Air Force "JTIDS Class 2 TDMA," June 1982 through December 1985 "JTIDS Class 2 TDMA" Army Supplements, June 1982 through December 1985 "JTIDS DTDMA" Navy, December 1984 through December 1990 "JTIDS Class 2 TDMA," and December 1991 through December 1995 "JTIDS" SARs. Development costs are tracked throughout the program's duration, with most production costs in the SARs related to the host platforms in which JTIDS terminals are to be integrated. The exception is in the June 1982 through December 1984 "JTIDS DTDMA" Navy SARs, in which projected Navy production costs are included. The Navy's DTDMA terminal was abandoned in 1985, when the Navy adopted the Air Force's TDMA terminal. The program has valid MS II and MS III baselines, but only development CGFs can be determined, due to absent production data.

Longbow Apache—Fire Control Radar

Data are from the December 1990, December 1991, and December 1993 and subsequent "Longbow Apache," and the December 1992 and June 1992 "Longbow" SARs. The airframe portion of the program is tracked separately. The MS II and MS III baselines are solid.

Over-the-Horizon Backscatter Radar (OTH-B) AN/FPS-118

Data are from the December 1983 through December 1989 "Over-the-Horizon Backscatter Radar (OTH-B)" and December 1990 "OTH-B" SARs. Units are operational sectors, which are heterogeneous because of specific site conditions accounting for up to 45 percent of the sector's cost. As a result, a 100-percent CIC is assumed in making quantity change adjustments. The MS II occurred before the first SAR program cost estimate; thus, the MS II baseline may not be suitable for database inclusion and analysis. The MS III baseline is solid.

Single-Channel Ground and Airborne Radio System–VHF (SINCGARS-V)

Data are from the December 1983 "Single Channel Ground and Airborne Radio System (SINCGARS-V)," December 1984 through December 1989 "Single Channel Ground and Airborne Radio System (SINCGARS)," and December 1990 through December 1999 "SINCGARS" SARs. The MS III baseline is solid.

Secure Mobile Antijam Reliable Tactical Terminal (SMART-T)

Data are from the December 1992 through December 2002 "SMART-T" SARs. The MS II and MS III baselines are solid.

Joint Tactical Communications (TRI-TAC)—Communication Nodal Control Element (CNCE) (AN/TSQ-111)

This is one of three cost tracks for the TRI-TAC program. Data are from the December 1984 through December 1988 "Joint Tactical Communications (TRI-TAC) Program" SARs. The MS III baseline is solid.

Satellites

Defense Meteorological Satellite Program (DMSP), Block 5D-2 Improved/5D-3

Data are from the December 1983 through December 1989 "DMSP Block 5D-2 Improved/5D-3" and December 1990 through December 1998 "DMSP" SARs. The MS III baseline is solid.

Defense Satellite Communication System (DSCS) Phase III

Data are from the June 1977 through September 1991 "DSCS III" SARs. The MS II and MS III baselines are solid.

Defense Support Program (DSP) 81

Data are from the December 1983 through December 1996 "DSP" SARs. Program funding began in 1967, but legacy DSP satellites and their costs in FY 1980 and prior are omitted from the SAR track. This SAR track begins with the third-generation DSP satellites, numbers 14 and subsequent (generation-one and -two satellites consisted of four developmental and nine production units). The MS III represents the beginning of production for the third-generation satellites, with the final of the previous generation funded eight years prior. The MS III baseline is solid.

GPS Navigational Satellite Timing and Ranging (NAVSTAR)—Initial Satellite Constellation Segments (Blocks I Through IIA)

This program portion covers the 12 developmental and first 28 production GPS satellites. User equipment and replacement and follow-on satellites are tracked separately. Data are from the December 1980 through December 1985 and December 1990 through December 1992 "NAVSTAR GPS," September 1986 through December 1988 "NAVSTAR GPS/User Equipment," and December 1989 "NAVSTAR GPS/Space and Control and User Equipment" SARs. The MS II and MS III baselines are solid.

Ultrahigh-Frequency (UHF) Follow-On Satellite

Data are from the December 1988 through December 1997 "UHF Follow-On" SARs. The program contains no development funding. The MS III baseline is solid.

Vehicles

M-1A2 Abrams Tank—Block II Improvement Model

Data are from the December 1992 through December 1994 and December 1998 through December 2003 "M1A2 Abrams Upgrade" and December 1995 through December 1997 "Abrams Upgrade" SARs. This program upgrades existing M1 tanks, with LRIP units coming directly off the M1A1 production line and all subsequent units coming from the field. In M1 SARs prior to December 1992, all M1A, M1A1, and M1A2 units were reported, but costs for the separate models were not identified separately. The December 1992 SAR excludes all M1A and M1A1 costs. For definitional consistency, costs are adjusted in the December 1999 through December 2003 SARs to remove costs associated with a preexisting retrofit program that was transferred to the SAR-reported funding beginning in December 1999. Although there was a documented MS II decision point for the M1A2, there was no corresponding cost estimate. The MS III baseline is solid.

Bradley Fighting Vehicle System A3 Upgrade

Data are from the December 1993 through December 2004 "BFVS A3 Upgrade" SARs. The program upgrades existing Bradley fighting vehicle systems—no new combat vehicles are built. The MS II and MS III baselines are solid.

Other

Common Strategic Rotary Launcher

This program designed and built mechanisms housed within bomb bays to accommodate the launching of internally carried cruise missiles. Data are from the December 1985 through December 1988 "Common Strategic Rotary Launcher" SARs. The MS III baseline is solid.

MK 50 Torpedo Advanced Antisubmarine Warfare Torpedo

Data are from the June 1983 through December 1994 "MK 50 Torpedo" SARs. The program has solid MS II and MS III baselines.

CAPTOR (Encapsulated Torpedo)—Mine Mk 60 Mod 0

Data are from the March 1976 through December 1983 "CAPTOR" SARs. The MS III is solid.

APPENDIX B

Ongoing Programs

Aircraft

C-17 Direct Delivery Airlift Aircraft—Globemaster III

Data are from the December 1985 through December 1989, December 1996 through December 1998, and December 2001 through December 2003 "C-17A," December 1990 through December 1995 and December 1999 "C-17," and December 2004 "C-17A Globemaster III" SARs. Data from the program's December 1983 through December 1984 SARs are omitted because they are from before the MS II and because an MS I baseline estimate cannot be established. The MS II and MS III baselines are solid.

F/A-18E/F Naval Strike Fighter

Data are from the June 1992 through December 2004 "F/A-18 E/F" SARs. The MS II and MS III baselines are solid.

F/A-22 Raptor Stealth Fighter

Data are from the December 1986 through December 1990 "ATF," December 1991 through December 2001 "F-22," and December 2002 through December 2004 "F/A-22 Raptor" SARs. The MS I, MS II, and MS III baselines are solid.

JPATS

Data are from the December 1996 through December 2004 "JPATS" SARs. The program MS II and MS III contracts were awarded just days apart; thus, no distinction can be made between a development and procurement baseline estimate. The MS II/III baseline is solid.

V-22 "Osprey" (JVX) Joint Service Advanced Vertical-Lift Aircraft

Data are from the December 1983 "JVX," December 1984 through December 1986 "V-22 (JVX)," and December 1987 through December 2004 "V-22 (OSPREY)" SARs. The production program was terminated in April 1989 and reinstated in August 1995. The MS I, MS II, and MS III baselines are solid.

Electronics

E-3 Sentry AWACS RSIP

Data are from the December 1989 "Radar System Improvement Program (RSIP)," the December 1990 through December 1992 "AWACS RSIP," and the December 1993 through December 2003 "E-3 AWACS RSIP" SARs. The program updates the aircraft's radar hardware and software. The MS II and MS III baselines are solid.

B-1B Conventional Mission Upgrade Program—Computer Upgrade

Data are from the December 1996 and December 1997 "B-1 CMUP-Computer" and December 1998 through December 2004 "B-1B CMUP" SARs. The program replaces six existing onboard computers with four new ones. The MS II and MS III baselines are solid.

Cooperative Engagement Capability

Data are from the September 1995 through December 2004 "CEC" SARs. The initial SAR and the date of the program's MS II baseline estimate in that SAR are after the program's development contract award; thus, the MS II baseline might not be suitable for database inclusion and analyses. The MS III baseline is solid.

GBS

Data are from the December 1997 through June 2005 "GBS" SARs. This is a near-worldwide, one-way information transmission system that utilizes the Navy UHF follow-on satellites for its space segment. The system consists of transmit suites, receive suites (two designs), injection points (two designs), and end-to-end system integration. Due to the heterogeneous unit mix, no quantity-adjusted CGFs are calculated. The MS II/III baseline is solid.

JDAM

Data are from the December 1994 through December 2004 "JDAM" SARs. The JDAM is a guidance kit that straps onto 2,000-, 1,000-, and 500-lb bombs. All three MSs are solid.

JSOW (formerly, Advanced Interdiction Weapon System [AIWS])—Baseline (AGM-154A)/BLU-108 Versions

Data are from the December 1991 "AIWS" and June 1992 through December 2004 "JSOW" and "JSOW (AIWS)" SARs. SAR cost data for these two versions are reported in aggregate; SAR schedule data are reported separately. The unitary version is reported and tracked separately. The MS I, MS II, and MS III baseline estimates are solid.

JSOW—Unitary Version

Data are from the December 1995 through December 2004 "JSOW" and "JSOW (AIWS)" SARs. Prior "JSOW" and "AIWS" SARs do not report data for the unitary version. The baseline and BLU-108 versions are reported and therefore tracked separately (in a single file for both). The MS II and MS III baseline estimates are solid.

Maneuver Control System (MCS)—Blocks IIIa Through IV

Data are from the December 1991 through December 2004 "MCS" SARs. Only the latter portion of the program (1990 and subsequent years for RDT&E; 1991 and subsequent years for procurement) is tracked. This latter portion includes what was originally known as the common hardware/software units/system and what is currently called the MCS blocks IIIa/b and IV. All program activity prior to 1990 is omitted from both the cost and schedule tracks. The program has solid MS II and MS III baselines.

MIDS-LVT

Data are from the December 1993 through December 2004 "MIDS-LVT" SARs. The MS II and MS III baselines are solid.

Minuteman III Guidance Replacement Program

Data are from the December 1993 through December 2003 "MMIII GPR–Phase I" and "MMIII GPR" and December 2004 "Minuteman III GRP" SARs. The MS II and MS III baselines are solid.

National Airspace System (NAS)

Data are from the December 1993 through June 2005 "NAS" SARs. This program modernizes radar approach control facilities in parallel with and to be compatible with Federal Aviation Administration facilities. At no point does this program reach the ACAT 1 value thresholds. Clear MS dates were difficult to determine because each of the four major subsystems within the whole has its own MS. The program does, however, have three solid MS baselines.

Helicopters

CH-47F "Chinook" Improved Cargo Helicopter (ICH)

Data are from the June 1998 "ICH," December 1998 "ICH (CH-47F)," and December 1999 through December 2004 "CH-47 (ICH)" SARs. This began as a service life extension program for the CH-47D and evolved into a mixture of mostly structural life extension programs for existing CH-47D and MH-47G helicopters, along with some new-build units. Most units involve structural modification and incorporation of an open electronic architecture. Engine replacement is not part of the shelf-life extension program: New engines are installed prior to each helicopter's introduction into the ICH program. The MS II and MS III baselines are solid.

H-1 USMC Helicopter Upgrade

Data are from the December 1996 through December 2002 "USMC H-1 Upgrades" and the December 2003 through December 2004 "H-1 Upgrades" SARs. The MS II and MS III baselines are solid.

Longbow Apache—Airframe

Data are from the December 1990, December 1991, and December 1993 through December 2004 "Longbow Apache," and the December 1992 and June 1992 "Longbow" SARs. The fire-control radar portion of the program is tracked separately. The MS II and MS III baselines are solid.

MH-60R Multimission Helicopter Upgrade

Formerly known as the SH-60R (or LAMPS Mark III, Block II upgrade). Data are from the December 1994 through September 2000 "SH-60R" and the September 2001 through December 2004 "MH-60R" SARs. Two years after LRIP approval, the program was changed from remanufacturing existing airframes to procuring new aircraft. The MS II contract was awarded prior to the estimate date in the first SAR. The MS II baseline estimate date is not consistent with the MS II commitment date and therefore may not be suitable for database inclusion and analysis. The MS III baseline is solid.

MH-60S Vertical Replenishment Helicopter

Formerly known as the CH-60S, this program's data are from the September 1998 through September 2000 "CH-60S" and September 2001 through June 2005 "MH-60S" SARs. The MS II and MS III baselines are solid.

Missiles

AIM-9X/Short-Range Air-to-Air Missile

Data are from the December 1994 through December 2004 "AIM-9X" SARs. The program's MS I, MS II, and MS III baselines are solid.

GLMRS (Formerly, Multiple-Launch Rocket System)

Data are from the December 1998 through December 2001 "MLRS Upgrade Program" and December 2002 through December 2004 "GMLRS" SARs. The launcher and extended-range rocket portions of the system are tracked separately. The MS II is solid.

JASSM

Data are from the September 1996 through December 2004 "JASSM" SARs. The MS I, MS II, and MS III baselines are solid.

Javelin AAWS-M

Data are from the September 1989 through December 1989 "Advanced Antitank Weapons System–Medium," December 1990 "AAWS-M (Medium)," September 1991 through December 1995 "Javelin (AAWS-M)," and December 1996 through December 2004 "Javelin" SARs. The MS II and MS III are solid.

Longbow Hellfire Antitank Air-to-Ground Missile

This is a subsystem of the AH-64 Apache Weapon System. Data are from the December 1990 through December 2004 "Longbow Hellfire" SARs. The MS II and MS III baselines are solid.

Patriot Guided Missile System, PAC-3 Missile Portion of the Air Defense PAC-3 System

Data are from the December 1994 through December 2003 "Patriot PAC-3" and the December 2004 "Patriot/Meads CAP" SARs. Prior to the initial Patriot PAC-3 SAR, the missile upgrade effort was known as the extended-range interceptor program. The fire-control unit portion of the program is tracked separately. The MS II and MS III are solid.

Tactical Tomahawk (RGM-109E/UGM-109E)

Data are from the December 1998, December 1999, and December 2003 "Tactical Tomahawk," September 2001 through December 2002

"Tomahawk (R/UGM-109)," and September 2004 through December 2004 "Tomahawk (R/UGM-109E)" SARs. The MS II and MS III are solid.

UGM-133A Trident D-5 Missile

Data are from the December 1983 through December 1989 "Trident II (D5) Missile" and December 1990 through December 2004 "Trident II Missile" SARs. Program costs exclude those related to nuclear capability. The program has solid MS II and MS III baselines.

Space

EELV

Data are from the December 1996 through June 2005 "EELV" SARs. The MS I and MS II/III are solid.

Minuteman III Propulsion Replacement Program (Minuteman III PRP) (LGM-30G)

Data are from the June 1996 through December 2004 "Minuteman III PRP" SARs. The MS II appears solid and is included in the database and associated analyses. The MS II occurred two years prior to the initial SAR, and the estimate in that SAR is identical to that of the MS. However, the SAR states that the MS estimate is dated June 30, 1994; thus, it is coincident with the MS II date. Apparently, the estimate was simply not updated in the two years prior to the initial SAR. The MS III is solid.

SBIRS-Hi Component

Data are from the December 1996 through December 2004 "SBIRS" SARs. SBIRS-Hi provides geosynchronous earth orbit satellites, highly elliptical orbit payloads hosted by other (external to the SBIRS-Hi program) satellites, and some ground portions of the overall SBIRS architecture. The MS II is solid.

Vehicles

FMTV

Data are from the December 1988 "Family of Medium Tactical Vehicles (FMTV)" and December 1989 through December 2004 "FMTV" SARs. The MS II and MS III baselines are solid.

APPENDIX C

Statistical Analysis of DCGF

This appendix statistically examines the trend of development cost growth.

Figure C.1 plots DCGF at five years past MS B over time. The trend seems to be a slight downward slope when disregarding the two outliers, that is, those points with DCGFs greater than or equal to five.

Figure C.1
DCGF at Five Years Past MS B

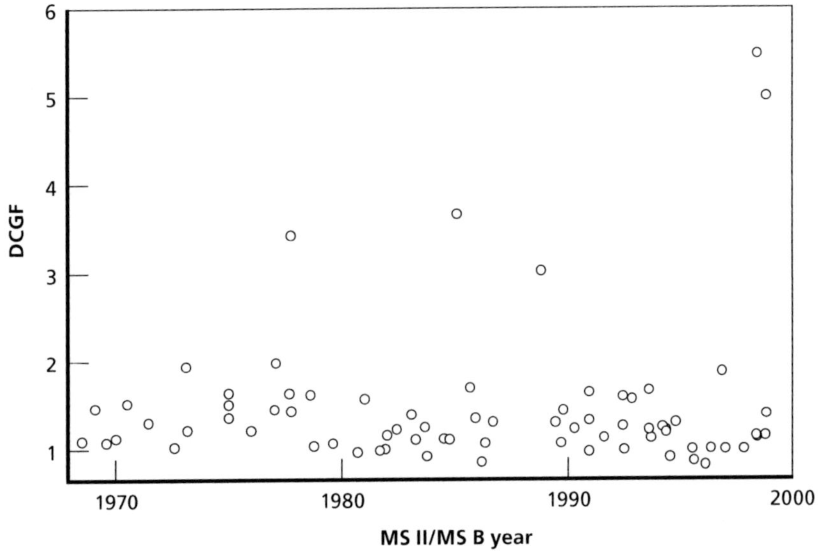

RAND *MG588-C.1*

Due to similarities of type categories, an alternate classification scheme is used; programs of similar types are classified into groups (e.g., cruise missiles; missile types become a missile group). Ten types become six groups. First, we will examine whether the effect of decade (time) on DCGF is significant even when adjusting for type of program. Then, we further examine whether a different inference is obtained when using groups as opposed to types in our statistical models.

Since Figure C.1 suggests that there is more variability in earlier decades than in the 1990s (when disregarding the two outliers), we will examine the variability of DCGF change over time.

Statistical Analysis

As indicated in Table C.1, which provides summary statistics of DCGF over time, median DCGF varies slightly over time, both with and without considering the two outliers. (Note that, for these statistical analyses, we operationalize time as decade categories for the summary statistics and for our regression models.) Median DCGF decreased from 1.43 in the 1970s to 1.14 in the 1980s and then rose to 1.2 in the 1990s. Table C.1 also indicates that the distribution of program type varies over time. For example, 80 percent of the cruise missile programs were developed in the 1970s. As we can see from model 2 in Table C.2, launch vehicles, cruise missiles, and vehicles have the highest DCGF among the various program types. Therefore, in answering the question of whether DCGFs have decreased over time, it is important to adjust for program type when looking at the relationship between time (decade) and DCGF as program types vary over time as well. We fit ordinary least squares (OLS) regression models to perform this adjustment.

Models 3 and 4 in Table C.2 are our full models, in which we look at the effect of time on DCGF while adjusting for program type in the model. In both models, we see that none of the coefficients for the 1980s and 1990s is statistically significant while adjusting for program type. It is important to note that these models are not very stable due to the small number of programs in some of the type categories.

Table C.1
Summary Statistics

	Decade					
DCGF Statistic	**1970s[a]**		**1980s**		**1990s**	
DCGF (n = 74)						
n	21		24		29	
Mean (std. deviation)	1.49 (0.53)		1.37 (0.65)		1.24 (0.29)	
Median	1.43		1.14		1.19	
DCGF (n = 76)						
n	21		24		31	
Mean (std. deviation)	1.49 (0.53)		1.37 (0.65)		1.50 (1.04)	
Median	1.43		1.14		1.21	
Program type (n = 76)	**n**	**(%)**	**n**	**(%)**	**n**	**(%)**
Aircraft	6	40	6	40	3	20
Cruise missiles	4	80	0	0	1	20
Electronic aircraft	3	60	2	40	0	0
Electronics	0	0	7	37	12	63
Helicopters	2	25	1	13	5[b]	63[b]
Launch vehicles	0	0	1	33	2	67
Missiles	4	29	5	36	5[b]	36[b]
Other	0	0	1	100	0	0
Satellites	2	67	0	0	1	33
Vehicles	0	0	1	33	2	67

NOTE: n = 76 represents the entire sample. n = 74 represents the entire sample minus the two outlier observations (DCGF ≥ 5).

[a] We consider three programs from the 1960s as occurring in the 1970s for this analysis.

[b] The two outliers are the GMLRS program, which is a missile program from the 1990s (DCGF = 5.00), and the MH-60S program, which is a helicopter from the 1990s (DCGF = 5.48).

For example, there is only one *other* program, and, in models 3 and 4, the standard error is almost eight times the magnitude of the coefficient for this program. Looking at the summary statistics of DCGF over time, DCGF varies by decade, but, after fitting a model and adjusting for program type, this difference between decades is not statistically significant.

We further collapse program types into groups and run the models, adjusting for these groups as a sensitivity analysis. These are models 5 and 6 in Table C.3. The decreased R-square and consistently larger standard errors relative to the regression coefficients indicate that these models do not fit as well as models 3 and 4. Since none of the p-values for the decade coefficient is significant at the $p=0.05$ level, we do not have any evidence of a time trend in DCGF after adjustment for program type.

In an additional sensitivity analysis, we further explored the relationship between time and DCGF by weighting each of the programs by their actual constant dollar budgets at five years into the program. Since the programs differ in size, weighting the regressions in Tables C.2 and C.3 by cost highlights any time trends in DCGF while giving higher-cost programs more emphasis in the regression equation. Tables C.4 and C.5 summarize our weighted OLS models. Looking at the regression coefficients for decade in models 3 through 6, none of the p-values is significant at the $p=0.05$ level. Therefore, we do not have any evidence of a trend in DCGF even after weighting by program size.

Another question of interest is whether there is greater variability of DCGF in earlier decades. To answer this question, we look at the residuals from our regressions in models 3 and 5. We compute the estimated variance of these residuals for each decade. From model 3, the estimated variances (of the residual term) for the 1970s, 1980s, and 1990s are 0.23, 0.22, and 0.13, respectively. From model 3, for the three comparisons (1970s versus 1980s, 1980s versus 1990s, and 1970s versus 1990s), the p-values from our F tests are 0.96, 0.21, and 0.20, respectively. Therefore, from model 3, we do not have any evidence that variability is significantly different across the three decades. From model 5, the estimated variances are 0.29, 0.26, and 0.12, respectively. From model 5, for the three comparisons (1970s versus 1980s, 1980s versus 1990s, and 1970s versus 1990s), the p-values from our F tests are 0.82, 0.06, and 0.04, respectively. Therefore, we do have some mild evidence, from one of our sensitivity models, that there is greater variability of DCGF in earlier decades compared to the 1990s when disregarding the two outliers.

Note that, in these analyses, we run our models both with and without the two outliers. Although those outliers are indeed valid observations, they exert a large influence in our regression models, and the models fit better when excluding them. If further data collection were to proceed in future decades and more observations with DCGFs greater than five were obtained, then these two observations that we currently consider "outliers" could contextually be placed as an actual "trend." But further data would be needed before making such a conclusion. We do have some evidence that there is greater variability in DCGF in earlier decades compared to the 1990s when not considering the two outliers. However, if we were to consider these two outliers, our conclusions on variability would be reversed; that is, variability would in fact be *greater* in the 1990s than in the earlier decades and these differences would be statistically significant ($p < 0.01$).

Table C.2
OLS Regression Models Using Program Type

Program Characteristic or Type	Model 1 Coeff.	Model 1 Std. error	Model 1 p-value	Model 2 Coeff.	Model 2 Std. error	Model 2 p-value	Model 3 Coeff.	Model 3 Std. error	Model 3 p-value	Model 4 Coeff.	Model 4 Std. error	Model 4 p-value
Model description	Time only			Type only			Type + time			Type + time		
n	74			74			74			76		
1970s				Reference category			Reference category			Reference category		
1980s	−0.12	0.15	0.43				−0.04	0.16	0.82	0.04	0.27	0.88
1990s	−0.25	0.14	0.09				−0.26	0.16	0.11	0.07	0.27	0.80
Aircraft				Reference category			Reference category			Reference category		
Cruise missile				0.60	0.24	0.02	0.58	0.25	0.02	0.61	0.43	0.16
Electronic aircraft				0.43	0.24	0.08	0.38	0.24	0.12	0.44	0.42	0.29
Electronics				0.04	0.16	0.81	0.15	0.17	0.39	0.01	0.30	0.97
Helicopter				0.26	0.22	0.24	0.34	0.22	0.12	0.75	0.37	0.05
Launch vehicle				0.76	0.30	0.01	0.87	0.30	0.01	0.72	0.52	0.17
Missile				0.07	0.18	0.69	0.10	0.18	0.58	0.33	0.30	0.28

Table C.2—Continued

Program Characteristic or Type	Model 1			Model 2			Model 3			Model 4		
	Coeff.	Std. error	p-value	Coeff.	Std. error	p-value	Coeff.	Std. error	p-value	Coeff.	Std. error	p-value
Other				0.09	0.49	0.85	0.06	0.49	0.90	0.08	0.85	0.92
Satellite				0.49	0.30	0.11	0.51	0.30	0.09	0.50	0.52	0.34
Vehicle				0.66	0.30	0.03	0.78	0.30	0.01	0.63	0.52	0.24
Intercept	1.48	0.11	< 0.001	1.16	0.12	< 0.001	1.22	0.15	< 0.001	1.12	0.26	< 0.001
R-square		0.04			0.23			0.27			0.13	

NOTE: Shading indicates statistical significance at the 5-percent level or less. n = 76 represents the entire sample. n = 74 represents the entire sample minus the two outlier observations (DCGF ≥ 5). The specification for each OLS model in the table is given by the rows in which coefficient information is provided. For example, the regression equation for model 2 is as follows:

$$\text{DCGF} = \beta_0 + \beta_1 \text{cruise missile} + \beta_2 \text{electronic aircraft}$$
$$+\beta_3 \text{electronics} + \beta_4 \text{helicopter} + \beta_5 \text{launch vehicle}$$
$$+\beta_6 \text{missile} + \beta_7 \text{other} + \beta_8 \text{satellite} + \beta_9 \text{vehicle} + \varepsilon.$$

The specification for model 1 is simply $\text{DCGF} = \beta_0 + \beta_1 \text{1980s} + \beta_2 \text{1990s} + \varepsilon$.

Table C.3
OLS Regression Models Using Program Group (Sensitivity Analyses)

	Model					
	5			6		
Program Characteristic or Type	Coeff.	Std. error	p-value	Coeff.	Std. error	p-value
Model Description	Group + time			Group + time		
n	74			76		
1970s	Reference category			Reference category		
1980s	−0.10	0.15	0.50	−0.00	0.26	0.99
1990s	−0.28	0.15	0.07	0.13	0.25	0.62
Aircraft + electronic aircraft + helicopters	Reference category			Reference category		
Electronics	0.00	0.16	0.99	−0.30	0.26	0.26
Cruise missiles + missiles	0.07	0.15	0.63	0.11	0.24	0.65
Other	−0.06	0.50	0.91	−0.16	0.84	0.85
Launch vehicles + satellites	0.53	0.22	0.02	0.30	0.37	0.42
Vehicles	0.63	0.30	0.04	0.31	0.50	0.54
Intercept	1.41	0.12	< 0.001	1.42	0.21	< 0.001
R-square	0.17			0.06		

NOTE: Shading indicates statistical significance at the 5-percent level or less. n = 76 represents the entire sample. n = 74 represents the entire sample minus the two outlier observations (DCGF ≥ 5).

Table C.4
Weighted OLS Regression Models Using Program Type

Program Characteristic or Type	Model											
	1			2			3			4		
	Coeff.	Std. error	p-value	Coeff.	Std. error	p-value	Coeff.	Std. error	p-value	Coeff.	Std. error	p-value
Model description	Time only			Type only			Type + time			Type + time		
n	74			74			74			76		
1970s												
1980s	−0.06	0.13	0.65				0.01	0.11	0.96	0.00	0.15	0.99
1990s	−0.11	0.13	0.40				−0.10	0.11	0.35	−0.05	0.15	0.73
Aircraft				Reference category			Reference category			Reference category		
Cruise missile				0.40	0.18	0.03	0.37	0.20	0.06	0.39	0.25	0.14
Electronic aircraft				0.44	0.17	0.01	0.40	0.18	0.03	0.42	0.23	0.07
Electronics				0.11	0.15	0.47	0.11	0.15	0.45	0.11	0.20	0.57
Helicopter				0.26	0.22	0.25	0.29	0.23	0.21	0.55	0.28	0.06
Launch vehicle				1.51	0.23	< 0.001	1.51	0.23	< 0.001	1.51	0.30	< 0.001

Table C.4—Continued

Program Characteristic or Type	Model											
	1			2			3			4		
	Coeff.	Std. error	p-value	Coeff.	Std. error	p-value	Coeff.	Std. error	p-value	Coeff.	Std. error	p-value
Missile				–0.11	0.12	0.36	–0.13	0.12	0.30	–0.06	0.16	0.72
Other				0.08	0.33	0.82	0.04	0.34	0.92	0.06	0.44	0.90
Satellite				0.52	0.19	< 0.01	0.56	0.19	< 0.01	0.54	0.25	0.03
Vehicle				0.35	0.46	0.44	0.40	0.46	0.38	0.38	0.60	0.53
Intercept	1.36	0.10	< 0.001	1.17	0.06	< 0.001	1.21	0.09	< 0.001	1.19	0.12	< 0.001
R-square		0.01			0.49			0.50			0.36	

NOTE: Shading indicates statistical significance at the 5-percent level or less. n = 76 represents the entire sample. n = 74 represents the entire sample minus the two outlier observations (DCGF ≥ 5).

Table C.5
Weighted OLS Regression Models Using Program Group (Sensitivity Analyses)

Program Characteristic or Type	Model 5 Coeff.	Model 5 Std. error	Model 5 p-value	Model 6 Coeff.	Model 6 Std. error	Model 6 p-value
Model description	Group + time			Group + time		
n	74			76		
1970s	Reference category			Reference category		
1980s	−0.06	0.12	0.58	−0.06	0.15	0.66
1990s	−0.21	0.12	0.08	−0.14	0.15	0.34
Aircraft + electronic aircraft + helicopter	Reference category			Reference category		
Electronics	0.09	0.17	0.61	0.06	0.21	0.77
Cruise missiles + missiles	−0.06	0.12	0.60	−0.02	0.15	0.87
Other	−0.01	0.38	0.98	−0.00	0.48	0.99
Launch vehicles + satellite	0.89	0.17	< 0.001	0.86	0.21	< 0.001
Vehicles	0.39	0.52	0.46	0.34	0.65	0.60
Intercept	1.32	0.09	< 0.001	1.32	0.11	< 0.001
R-square	0.32			0.20		

NOTE: Shading indicates statistical significance at the 5-percent level or less. n = 76 represents the entire sample. n = 74 represents the entire sample minus the two outlier observations (DCGF ≥ 5).

APPENDIX D

Statistical Comparison of Military Service DCGF

In Appendix C, we used statistical methods to explore whether different weightings for weapon system types affect our conclusions about whether there is a time trend for cost growth. We found that there were no significant trends. However, one question remains as to whether the service (i.e., Air Force, Army, or Navy) programs have different cost growth trends. In this appendix, we statistically examine whether the three services differ regarding their levels of development cost growth for their respective programs. Again, we focus on the development cost growth five years after MS B.

Table D.1 provides descriptive statistics (mean, median, and standard deviation) of DCGF by service for each decade. Each of the 76 ongoing and completed programs belongs to one of the three services: Air Force, Army, or Navy. Examining the means and medians for all decades combined, we see that the services do not differ greatly in their levels of DCGF. Both with and without the two outliers, the services' DCGFs are similar.

Since both program type and decade (time) are hypothesized to influence DCGF, we fit regression models for service and adjust for program type and decade. Table D.2 contains the five OLS regressions that were fitted. For simplicity, we present only the statistics for the coefficients related to the service, as we are primarily concerned about inference of the service variables. The first regression, model 1, contains indicator variables for only the Army and the Navy (the Air Force is the reference group). As we have observed before, neither of the coefficients is statistically significant.

The remaining four regressions in Table D.2 parallel models 3 through 6 in Appendix C. Models 3 through 6 in Table D.2 include Army and Navy indicators and adjust for program type (or program group) and decade. We see in all these models that the coefficients for the Army and Navy are close to zero and not significant. Due to this lack of statistical significance, we conclude that the respective programs of the three services do not differ significantly in their DCGF levels.

In an additional sensitivity analysis, we further explored the relationship between service and DCGF by weighting each of the programs by their actual budgets in constant dollars at five years into the program. Since the programs differ in size, weighting the regressions in Table D.2 by cost should highlight whether there are any associations between service and DCGF while simultaneously giving higher-cost programs more emphasis in the regression equation. Table D.3 summarizes our weighted OLS models. The regression coefficients and associated p-values for the services in models 1 and 3 through 6 show that, when looking at service alone, the weighted regression suggests that Navy programs have significantly lower DCGFs than do Air Force programs. However, after adjusting for time and program type, the p-values for Army and Navy programs are not significant at the $p=0.05$ level. Therefore, we do not have any evidence of differential levels of DCGF by service even after weighting by program size.

Table D.1
DCGF Summary Statistics by Service

	Decade															
	1970s[a]				1980s				1990s				All Decades			
DCGF Statistic	Mean	Std. Dev.	n	Median	Mean	Std. Dev.	n	Median	Mean	Std. Dev.	n	Median	Mean	Std. Dev.	n	Median
Service (n = 76)																
Air Force	1.57	0.67	12	1.40	1.47	0.77	11	1.22	1.15	0.29	11	1.12	1.40	0.62	34	1.19
Army	1.03	n/a	1	1.03	1.49	0.79	6	1.30	1.66[b]	1.19	10	1.27	1.56	1.01	17	1.30
Navy	1.41	0.19	8	1.46	1.10	0.10	7	1.11	1.71[b]	1.36	10	1.26	1.44	0.88	25	1.25
Service (n = 74)																
Air Force	1.57	0.67	12	1.40	1.47	0.77	11	1.22	1.15	0.29	11	1.12	1.40	0.62	34	1.19
Army	1.03	n/a	1	1.03	1.49	0.79	6	1.30	1.29	0.21	9	1.23	1.35	0.50	16	1.27
Navy	1.41	0.19	8	1.46	1.10	0.10	7	1.11	1.29	0.35	9	1.25	1.28	0.27	24	1.23

NOTE: n = 76 represents the entire sample. n = 74 represents the entire sample minus the two outlier observations (DCGF ≥ 5).
n/a = not applicable.

[a] We classify three programs from the 1960s as occurring in the 1970s in this analysis.

[b] The two outliers are the Army's GMLRS program, which is a missile program from the 1990s (DCGF = 5.00) and the Navy's MH-60S program, which is a helicopter program from the 1990s (DCGF = 5.48).

Table D.2
OLS Regression Models

	Model														
	1			3			4			5			6		
Model Statistic	**Coeff.**	**Std. error**	**p-value**	**Coeff.**	**Std. error**	**p-value**	**Coeff.**	**Std. error**	**p-value**	**Coeff.**	**Std. error**	**p-value**	**Coeff.**	**Std. error**	**p-value**
Model description[a]	Service only			Service + type + time			Service + type + time			Service + group + time			Service + group + time		
n	74			74			76			74			76		
Air Force	Reference category			Reference category			Reference category			Reference category			Reference category		
Army	−0.06	0.15	0.72	−0.05	0.19	0.79	−0.05	0.32	0.89	−0.02	0.17	0.89	0.11	0.29	0.71
Navy	−0.13	0.13	0.35	−0.04	0.14	0.78	0.01	0.25	0.98	−0.03	0.14	0.81	0.04	0.24	0.85
R-square	0.01			0.27			0.13			0.17			0.06		

NOTE: This table presents coefficients for two services: the Army and the Navy (the Air Force is the reference group). The numbering of the models 3 through 6 follows the model numbers in Appendix C, Tables C.2 and C.3. n = 76 represents the entire sample. n = 74 represents the entire sample minus the two outlier observations (DCGF ≥ 5).

[a] The specification for each OLS model is given by the model description. For example, the regression equation for model 1 is as follows: $\text{DCGF} = \beta_0 + \beta_1\text{Army} + \beta_2\text{Navy}$. The regression equation for model 4 is

$$\begin{aligned}\text{DCGF} &= \beta_0 + \beta_1\text{cruise missile} + \beta_2\text{electronic aircraft} + \beta_3\text{electronics}\\ &+\beta_4\text{helicopter} + \beta_5\text{launch vehicle} + \beta_6\text{missile} + \beta_7\text{other} + \beta_8\text{satellite}\\ &+\beta_9\text{vehicle} + \beta_{10}\,\text{1980s} + \beta_{11}\,\text{1990s} + \beta_{12}\text{Army} + \beta_{13}\text{Navy} + \varepsilon.\end{aligned}$$

Table D.3
Weighted OLS Regression Models (Service Coefficients Only)

	Model														
	1			3			4			5			6		
Model Statistic	Coeff.	Std. error	p-value	Coeff.	Std. error	p-value	Coeff.	Std. error	p-value	Coeff.	Std. error	p-value	Coeff.	Std. error	p-value
Model description	Service only			Service + type + time			Service + type + time			Service + group + time			Service + group + time		
n	74			74			76			74			76		
Air Force	Reference category			Reference category			Reference category			Reference category			Reference category		
Army	−0.15	0.20	0.45	−0.02	0.20	0.90	0.01	0.25	0.96	−0.01	0.20	0.94	0.10	0.24	0.68
Navy	−0.24	0.11	0.03	−0.09	0.10	0.35	−0.08	0.13	0.54	−0.13	0.11	0.21	−0.10	0.13	0.45
R-square	0.06			0.50			0.36			0.33			0.22		

NOTE: n = 76 represents the entire sample. n = 74 represents the entire sample minus the two outlier observations (DCGF ≥ 5).

Bibliography

Arena, Mark V., Irv Blickstein, Obaid Younossi, and Clifford A. Grammich, *Why Has the Cost of Navy Ships Risen? A Macroscopic Examination of the Trends in U.S. Naval Ship Costs Over the Past Several Decades*, Santa Monica, Calif.: RAND Corporation, MG-484-NAVY, 2006. As of January 16, 2007: http://www.rand.org/pubs/monographs/MG484/

Arena, Mark V., Robert S. Leonard, Sheila E. Murray, and Obaid Younossi, *Historical Cost Growth of Completed Weapon System Programs*, Santa Monica, Calif.: RAND Corporation, TR-343-AF, 2006. As of January 15, 2007: http://www.rand.org/pubs/technical_reports/TR343/

Arena, Mark V., Obaid Younossi, Lionel A. Galway, Bernard Fox, John C. Graser, Jerry M. Sollinger, Felicia Wu, and Carolyn Wong, *Impossible Certainty: Cost Risk Analysis for Air Force Systems*, Santa Monica, Calif.: RAND Corporation, MG-415-AF, 2006. As of January 15, 2007: http://www.rand.org/pubs/monographs/MG415/

Asher, Norman J., and Theodore F. Maggelet, "On Estimating the Cost Growth of Weapon Systems," Institute for Defense Analyses, Paper P-1494, September 1984.

Bruno, Michael, "'Bow Wave' of Acquisition Costs Coming, Rumsfeld Tells Senators," *Aerospace Daily and Defense Report*, April 28, 2005.

Calcutt, Harry M., Jr., *Cost Growth in DoD Major Programs: A Historical Perspective*, Washington, D.C.: Industrial College of the Armed Forces, Ft. McNair, 1993. As of January 16, 2007: http://stinet.dtic.mil/cgi-bin/GetTRDoc?AD=ADA276950&Location=U2&doc=GetTRDoc.pdf

Cook, Cynthia R., and John C. Graser, *Military Airframe Acquisition Costs: The Effects of Lean Manufacturing*, Santa Monica, Calif.: RAND Corporation, MR-1325-AF, 2001. As of January 15, 2007: http://www.rand.org/pubs/monograph_reports/MR1325/

Cottrill, Adrian, "Taking on a Cult of Mediocrity," *Upstream*, May 23, 2003.

Drezner, Jeffrey A., Jeanne M. Jarvaise, Ron W. Hess, Paul G. Hough, and Daniel M. Norton, *An Analysis of Weapon System Cost Growth*, Santa Monica, Calif.: RAND Corporation, MR-291-AF, 1993. As of January 15, 2007: http://www.rand.org/pubs/monograph_reports/MR291/

Fisher, Gene H., *Cost Considerations in Systems Analysis*, New York: American Elsevier, 1971.

Flyvbjerg, Bent, Nils Bruzelius, and Werner Rothengatter, *Megaprojects and Risks: An Anatomy of Ambition*, Cambridge, UK: Cambridge University Press, 2003.

Flyvbjerg, Bent, Mette Skamris Holm, and Soren L. Buhl, "Underestimating Costs in Public Works Projects: Error or Lie?" *Journal of the American Planning Association*, Vol. 68, No. 3, Summer 2002, pp. 279–295.

Fox, Bernard, Michael Boito, John C. Graser, and Obaid Younossi, *Test and Evaluation Trends and Costs for Aircraft and Guided Weapons*, Santa Monica, Calif.: RAND Corporation, MG-109-AF, 2004. As of January 15, 2007: http://www.rand.org/pubs/monographs/MG109/

Hanks, Christopher H., Elliot I. Axelband, Shuna Lindsay, Mohammed Rehand Malik, and Brett D. Steele, *Reexamining Military Acquisition Reform: Are We There Yet?*, Santa Monica, Calif.: RAND Corporation, 2005. As of March 5, 2007: http://www.rand.org/pubs/monographs/MG291/

Hedvall, Maria, "Change as a Cost Driver in Defence Procurement," *Defence and Peace Economics*, Vol. 15, No. 1, February 2004, pp. 101–108.

Hough, Paul G., "Pitfalls in Calculating Cost Growth from Selected Acquisition Reports," Santa Monica, Calif.: RAND Corporation, N-3136-AF, 1992. As of January 15, 2007: http://www.rand.org/pubs/notes/N3136/

Karp, Jonathon, "Defense Industry Braces for Report on Cost Overruns," *Wall Street Journal*, April 7, 2006a, p. A4.

———, "With New Rules, Weapons Projects Set Higher Costs," *Wall Street Journal*, April 10, 2006b, p. A6.

Karp, Jonathan, and Andy Pasztor, "A Firebrand Singes Defense Establishment; Speaking Harshly and Carrying a Big Stick, McCain Takes on Pentagon Procurement," *Wall Street Journal*, April 18, 2005, p. A4.

Kirkpatrick, David L. I., "Trends in the Costs of Weapon Systems and the Consequences," *Defence and Peace Economics*, Vol. 15, No. 3, June 2004, pp. 259–273.

Lorell, Mark A., and John C. Graser, *An Overview of Acquisition Reform Cost Savings Estimates*, Santa Monica, Calif.: RAND Corporation, MR-1329-AF, 2001. As of January 15, 2007: http://www.rand.org/pubs/monograph_reports/MR1329/

Lorell, Mark A., John C. Graser, and Cynthia R. Cook, *Price-Based Acquisition: Issues and Challenges for Defense Department Procurement of Weapon Systems*, Santa Monica, Calif.: RAND Corporation, MG-337-AF, 2005. As of January 15, 2007: http://www.rand.org/pubs/monographs/MG337/

Lorell, Mark A., Julia F. Lowell, and Obaid Younossi, *Evolutionary Acquisition: Implementation Challenges for Defense Space Systems*, Santa Monica, Calif.: RAND Corporation, MG-431-AF, 2006. As of January 15, 2007: http://www.rand.org/pubs/monographs/MG431/

Marshall, A. W., and W. H. Meckling, *Predictability of the Costs, Time and Success of Development*, Santa Monica, Calif.: RAND Corporation, P-1821, 1959.

Massachusetts Turnpike Authority, "The Big Dig," undated Web page. As of January 16, 2007: http://www.masspike.com/bigdig/

McNichol, D. L., "Growth in the Costs of Major Weapon Procurement Programs," Alexandria, Va.: Institute for Defense Analyses, Paper P-3832, 2004.

Merrow, Edward W., Lorraine M. McDonnell, and R. Ylmaz Arguden, *Understanding the Outcomes of Mega-Projects: A Quantitative Analysis of Very Large Civilian Projects*, Santa Monica, Calif.: RAND Corporation, R-3560-PSSP, 1988. As of January 16, 2007: http://www.rand.org/pubs/reports/R3560/

Merrow, Edward W., Kenneth E. Phillips, and Christopher W. Myers, *Understanding Cost Growth and Performance Shortfalls in Pioneer Process Plants*, Santa Monica, Calif.: RAND Corporation, R-2569-DOE, 1981. As of January 16, 2007: http://www.rand.org/pubs/reports/R2569/

Morgan, M. Granger, and Max Henrion, *Uncertainty: A Guide to Dealing with Uncertainty in Quantitative Risk and Policy Analysis*, Cambridge, UK: Cambridge University Press, 1990.

Novick, David, *Are Cost Overruns a Military-Industry-Complex Specialty?*, Santa Monica, Calif.: RAND Corporation, P-4311, 1970.

Pfleeger, Shari Lawrence, Felicia Wu, and Rosalind Lewis, *Software Cost Estimation and Sizing Methods: Issues and Guidelines*, Santa Monica, Calif.: RAND Corporation, MG-269-AF, 2005. As of January 15, 2007: http://www.rand.org/pubs/monographs/MG269/

Rich, Michael D., and Edmund Dews, *Thoughts on Reforming the Military Acquisition Process*, Santa Monica, Calif.: RAND Corporation, P-7352, 1987. As of January 16, 2007: http://www.rand.org/pubs/papers/P7352/

Rich, Michael D., Edmund Dews, and C. L. Batton, Jr., *Improving the Military Acquisition Process: Lessons from RAND Research*, Santa Monica, Calif.: RAND Corporation, R-3373-AF/RC, 1986. As of January 16, 2007:
http://www.rand.org/pubs/reports/R3373/

Sevastopulo, Demetri, "Cost of Jet Could Overrun by Dollars 5bn," *Financial Times*, April 8, 2005, p. 12.

Singer, Neil M., *Cost Growth in Weapon Systems: Recent Experience and Possible Remedies*, Washington, D.C.: Congressional Budget Office, October 12, 1982. As of January 16, 2007:
http://www.cbo.gov/ftpdoc.cfm?index=5102&type=1

Stem, David E., Michael Boito, and Obaid Younossi, *Systems Engineering and Program Management: Trends and Costs for Aircraft and Guided Weapons Programs*, Santa Monica, Calif.: RAND Corporation, MG-413-AF, 2006. As of January 15, 2007:
http://www.rand.org/pubs/monographs/MG413/

Tyson, K. W., B. R. Harmon, and D. M. Utech, "Understanding Cost and Schedule Growth in Acquisition Programs," Alexandria, Va.: Institute for Defense Analysis, Paper P-2967, 1994.

Tyson, K. W., J. R. Nelson, N. I. Om, and P. R. Palmer, "Acquiring Major Systems: Cost and Schedule Trends and Acquisition Initiative Effectiveness," Alexandria, Va.: Institute for Defense Analyses, Paper P-2201, 1998.

USC—*see* U.S. Code.

U.S. Code, Title 10, Section 2432, Selected Acquisition Reports, January 19, 2004.

U.S. Department of Defense, Department of Defense Directive (DoDD) 5000.1, The Defense Acquisition Program, October 23, 2000.

———, Department of Defense Instruction (DoDI) 5000.2, Operation of the Defense Acquisition Program, May 12, 2003.

———, "Nunn-McCurdy (NM) Unit Cost Breaches," 2002, supplemental materials for "Under Secretary of Defense Aldridge Roundtable," transcript, May 2, 2002. As of April 3, 2007 (follow link for "Nunn-McCurdy explanation and certification documents"):
http://www.defenselink.mil/transcripts/transcript.aspx?transcriptid=3425

———, Office of the Inspector General, *Acquisition: Major Defense Acquisition Programs Cost Growth*, Washington, D.C., Report No. D-2002-054, February 26, 2002. As of January 16, 2007:
http://www.dodig.mil/Audit/reports/fy02/02-054.pdf

U.S. General Accounting Office (now U.S. Government Accountability Office), *NASA Program Costs: Space Missions Require Substantially More Funding Than Initially Estimated*, Washington, D.C., GAO/NSIAD-93-97, December 1992. As of January 16, 2007:
http://archive.gao.gov/d36t11/148471.pdf

———, *NASA: Lack of Disciplined Cost-Estimating Processes Hinders Effective Program Management*, Washington, D.C., GAO-04-642, May 2004. As of January 16, 2007:
http://www.gao.gov/new.items/d04642.pdf

U.S. Government Accountability Office, *Defense Acquisitions: Assessments of Selected Major Weapon Programs*, Washington, D.C., GAO-05-301, March 2005. As of January 15, 2007:
http://www.gao.gov/new.items/d05301.pdf

Virginia Department of Transportation, "Springfield Interchange Improvement Project," undated Web page. As of January 16, 2007:
http://www.springfieldinterchange.com/

Younossi, Obaid, Mark V. Arena, Richard M. Moore, Mark A. Lorell, Joanna Mason, and John C. Graser, *Military Jet Engine Acquisition: Technology Basics and Cost-Estimating Methodology*, Santa Monica, Calif.: RAND Corporation, MR-1596-AF, 2003. As of January 15, 2007:
http://www.rand.org/pubs/monograph_reports/MR1596/

Younossi, Obaid, Michael Kennedy, and John C. Graser, *Military Airframe Costs: The Effects of Advanced Materials and Manufacturing Processes*, Santa Monica, Calif.: RAND Corporation, MR-1370-AF, 2001. As of January 15, 2007:
http://www.rand.org/pubs/monograph_reports/MR1370/

Younossi, Obaid, David E. Stem, Mark A. Lorell, and Frances M. Lussier, *Lessons Learned from the F/A-22 and F/A-18E/F Development Programs*, Santa Monica, Calif.: RAND Corporation, MG-276-AF, 2005. As of January 15, 2007:
http://www.rand.org/pubs/monographs/MG276/